Green Information and Communication Systems for a Sustainable Future

Green Engineering and Technology: Concepts and Applications

Series Editors
Brujo Kishore Mishra
GIET University, India
Raghvendra Kumar
LNCT College, India

The environment is an important issue these days for the whole world. Different strategies and technologies are being used to contribute to saving the environment. Technology is the application of knowledge to solve practical problems. Green technologies encompass various aspects of technology, which help us to reduce the human impact on the environment and to create ways of achieving sustainable development. Focusing on social justice, this book series will enlighten the green technology field in different ways, aspects, and methods. This technology helps people to understand the different ways of using resources to fulfil society's needs and demands. Some points will be discussed as the combination of involuntary approaches or government incentives, while a comprehensive regulatory framework will encourage the diffusion of green technology to the least-developed countries and to developing states or small islands requiring unique supports and measures to promote the green technologies.

Green Innovation, Sustainable Development, and the Circular Economy
Edited by Nitin Kumar Singh, Siddhartha Pandey, Himanshu Sharma, and Sunkulp Goel

Green Automation for Sustainable Environment
Edited by Sherin Zafar, Mohd Abdul Ahad, M. Afshar Alam, Kashish Ara Shakeel

AI in Manufacturing and Green Technology
Methods and Applications

Edited by Sambit Kumar Mishra, Zdzislaw Polkowski, Samarjeet Borah, and Ritesh Dash

Green Information and Communication Systems for a Sustainable Future
Edited by Rajshree Srivastava, Sandeep Kautish, and Rajeev Tiwari

For more information about this series, please visit: https://www.routledge.com/Green-Engineering-and-Technology-Concepts-and-Applications/book-series/CRCGETCA

Green Information and Communication Systems for a Sustainable Future

Edited by
Rajshree Srivastava
Sandeep Kautish
Rajeev Tiwari

CRC Press
Taylor & Francis Group
Boca Raton London New York

CRC Press is an imprint of the
Taylor & Francis Group, an **informa** business

First edition published 2020
by CRC Press
6000 Broken Sound Parkway NW, Suite 300, Boca Raton, FL 33487-2742

and by CRC Press
2 Park Square, Milton Park, Abingdon, Oxon, OX14 4RN

© 2021 Taylor & Francis Group, LLC

CRC Press is an imprint of Taylor & Francis Group, LLC

Library of Congress Cataloging-in-Publication Data

Names: Srivastava, Rajshree, editor. I Kautish, Sandeep, editor. I Tiwari, Rajeev, editor.
Title: Green information and communication systems for a sustainable future / edited by Rajshree Srivastava, Sandeep Kautish, and Rajeev Tiwari.
Description: First edition. I Boca Raton : CRC Press, 2021. I Series: Green engineering and technology I Includes bibliographical references and index.
Identifiers: LCCN 2020022758 (print) I LCCN 2020022759 (ebook) I ISBN 9780367894658 (hardback) I ISBN 9781003032458 (ebook)
Subjects: LCSH: Telecommunication systems--Environmental aspects.
Classification: LCC TK5105.5 .G728 2021 (print) I LCC TK5105.5 (ebook) I DDC 384.3--dc23
LC record available at https://lccn.loc.gov/2020022758
LC ebook record available at https://lccn.loc.gov/2020022759

ISBN: 978-0-367-89465-8 (hbk)
ISBN: 978-1-003-03245-8 (ebk)

Typeset in Times
by Deanta Global Publishing Services, Chennai, India

Contents

Preface

Nowadays, green computing is one of the major challenges for most IT organizations that involve medium- and large-scale distributed infrastructures like grids, clouds, and clusters. This book will focus on solutions for all aspects of green computing issues, such as energy efficiency, carbon footprint reduction, and temperature control management. The trade-offs between energy efficiency and performance have become key challenges that must be addressed in both distributed and traditional performance-oriented infrastructures. Particularly relevant is the so-called smart grid technology, seeking to optimize electricity distribution, especially electricity generated from renewable sources, and to promote the use of smart devices (including smart homes and autonomous cars), that require further research on distributed communications, energy storage, and the integration of various sources of energy.

The transition to a green economy poses many challenges; however, it also opens up significant opportunities for economic development in developed and developing countries. Through innovation in low-carbon technologies, countries can capture economic value 'at home' through entrepreneurship, job creation, and new venture development. These development goals can be simultaneously pursued while securing energy independence and mitigating the effects of climate change, as developing countries play a crucial role in catalyzing and deploying the clean technology innovations of the future.

This book compromises 14 well-structured chapters. Chapter 1 describes a survey of different localization algorithms, based on optimization methods, along with possible future directions. The two different scenarios, namely static and dynamic approaches to wireless sensor networks (WSNs), are also discussed. Chapter 2 surveys the literature on data aggregation techniques in WSNs. The literature suggests some unique ideas, covering most of the areas of WSN, such as clustering, data aggregation, mobile agent, and Trust-aware. This survey is intended for all who work with WSNs and in related fields, especially for aspiring researchers, to keeping up to date with progress in this field. One of the current topics in WSN is Trust-aware, which needs to be more fully explored. In Trust-aware, any malicious node must be detected by calculating the trust value, before running data aggregation processes. This approach is preferred over traditional cryptography methods because it has minimal computational cost.

Chapter 3 discusses how smart signal systems evolved, the current improvements to solve various problems, the further advances needed to improve the system, and potential research projects. Data mining principles have been used for a number of applications in banking nowadays. In Chapter 4, data mining principles have been used to assess security concerns and to develop better security solutions in e-banking. A dataset has been prepared by collecting information on perceptions of e-banking security from banking and e-banking customers through a questionnaire. The clustering technique has been used to detect statistically significantly different responses from the two groups of respondents, using Student's t-test and SPSS statistical software. Finally, the conclusion and future scope of the work has been discussed.

Chapter 5 addresses avenues for electric power production through renewable energy sources (RES), especially under the influence of varying climatic conditions. The Internet of things (IoT) has attracted huge popularity in recent times. IoT has been labelled as "The next Industrial Revolution" and is spreading across wireless sensor networks, where actual objects are linked *via* a mutual platform, the internet. The IoT paradigm is flagging its importance for finding the solution to problems in various areas. To simplify the upcoming work and to assist in the building of a prototype from the different models, Chapter 6 provides a detailed study of layer-defined, domain-defined and industry-defined IoT architectures. Major vendors like Cisco have also emphasized the progress needed to build new prototypes that normalize the complex IoT industry by releasing IoT platform architectures. This chapter also focuses on the core IoT, its basic elements, and the organized classification of various IoT architectures.

Chapter 7 deals with designing different modes of hybrid energy systems in order to understand their operation under various variable constraints. Chapter 8 describes the operation of a 5000 MVA Wind Electric Power Generation (WEPG) model to assess the various Power Quality (PQ) aspects. The need for a Maximum Power Point Tracking (MPPT) technique was also highlighted. Normally, a wind turbine operates under variable atmospheric conditions, and its operation is explained by the use of various types of generators and associated controllers. Since the integration of any RES stresses the need for PQ evaluation, the importance of Flexible AC Transmission Systems (FACTS) devices was explained in the mitigation of various PQ issues. The role of various FACTS devices was elaborated, with the importance of Unified Power Flow Controllers (UPFC) being highlighted. The capability of UPFC to mitigate various PQ issues and improve the available real power output of a wind generator was described, along with a detailed comparison of available FACTS controllers. Finally, the role of UPFC in compensating for the reactive power was explained. It was shown that UPFC can greatly enhance the performance of a wind generator, especially during the occurrence of non-symmetrical faults.

Chapter 9 deals with the assessment of energy systems and associated major services, which enhance efficient electrical energy generation and its delivery to the consumer on demand. In Chapter 10, PQ analysis, based on applying modulating techniques, can ensure the provision of harmonic-less power to medical and health care sector. The absence of such modulating techniques for controlling TIM drives in healthcare facilities can put the lives of patients at risk. The ultimate objective of the proposed systems is to deliver sinusoidal power at a constant magnitude. Due to harmonics presented in the electric power supply, many complexities, involving harmonics and electro-magnetic interference, are produced. Therefore, the proposed modulating systems have attempted to reduce the overall harmonics presented in line current and line voltages at the point of common coupling. Harmonic analysis has been carried out for the three-phase SQIM, using a power and harmonics analyzer (PHA) under dynamic loading conditions. It was shown that 3rd and 5th harmonics predominate. A passive filter was designed to improve PQ. By using a shunt passive filter, 3rd and 5th harmonics were reduced. Furthermore, pulse-width modulation (PWM) and discrete space vector pulse-width modulation (SVPWM) techniques are also used to mitigate the current harmonics. Both modulation techniques are coordinated with an AC-DC-AC converter. Observations reveal that both modulation techniques were effective, although SVPWM was superior to the PWM technique. The PHA is capable of measuring harmonics up to the 99th order, with a display of 50 harmonics on one screen. Further future work is also discussed.

Chapter 11 describes the concept of the smart home, its emergence and definition, and its components in details. The adoption of IoT in automation is also explained, with design strategies. Various issues and challenges of the smart home with different areas of use are also being discussed in detail. The automotive industry is now also known as the mobility industry and has become the backbone of the commute of every second person. Enormous amounts of research are being carried out to improve mobility, but very few studies are focusing on mobility in disaster-prone areas, where improving mobility under such conditions is of the utmost priority. Chapter 12 covers the mobility of vehicles in disaster-prone areas, especially with the focus of rescuing the vehicle in a post-disaster scenario, if it is trapped and its location is unknown. The unpredictable changes in climate, which can lead to disaster, result in the failure of cellular network systems. In such scenarios, the first 72 hours are crucial in terms of rescue operations, which prioritizes the design of a sustainable wireless communication network. This chapter presents progress in terms of design and development of an indigenous sensor network, with a focus on localization of vehicles in disaster-prone areas. The performance of the proposed system was tested in an open seismic area (OSA) and step-wise procedures are described to show how the indigenous network coped. The communication network for the present study consists of mobile wireless nodes which were interconnected with each other and incorporated the hybrid design of uni-lateration and fingerprinting algorithms. The chapter also discusses the design of an algorithm for localization and its subsequent validation. This system could result in incremental increases in safety and mobility in disaster-prone areas by monitoring and localizing the movement of vehicles with minimal environmental effects. Chapter 13 describes an offline payment system, which has the advantage that the payment and check balance are linkable. Through this, the passenger will be able to book a ticket, and the details will be sent to the conductor. The conductor will have the ID of the passenger sending the ticket, and the amount paid. After this, a verification message is displayed on both the passenger's and the conductor's screens. This system would be particularly appropriate for people dependent on public transport in isolated areas, where internet connection is weak or non-existent.

Microbial Fuel Cell (MFC) is a bioreactor which converts chemical energy present in bio-convertible substrates directly into electricity by the action of specific micro-organisms. These micro-organisms facilitate the conversion of the substrate molecules directly into electrons. The electrons released are collected to maintain production of electrical energy, proving to be an efficient source of sustainable energy production. This energy production is a result of oxidation of organic sources by anaerobic digestion process, with a by-product of bioelectricity production. MFC is an incredible technology with the capability to digest a wide range of substrates, configurations, and materials with bacteria to achieve the generation of bioelectricity, despite the fact that power levels are low. The two most significant parameters involved in MFC technology are power density and coulumbic efficiency. These parameters need to be optimized to achieve sustainable long-term power applications. Although the efficiency of MFC in electric power generation had been less in initial research, recent modifications have augmented the electric power output for various applications. Chapter 14 describes and discusses MFC power generation systems, which are capable of harnessing electric power with improved power quality levels.

This book gives a complete guide to the fundamental concepts, applications, algorithms, protocols, new trends, and challenges and research findings in the area of green information and communication systems. It is an invaluable resource, providing up-to-date information on the core and specialized issues in the field, and making it highly appropriate for both the novice and the experienced researcher in this field.

April, 2020
Rajshree Srivastava
Sandeep Kautish
Rajeev Tiwari

MATLAB® is a registered trademark of The MathWorks, Inc. For product information, please contact:

The MathWorks, Inc.
3 Apple Hill Drive
Natick, MA 01760-2098 USA
Tel: 508 647 7000
Fax: 508-647-7001
E-mail: info@mathworks.com
Web: www.mathworks.com

Editors

Rajshree Srivastava is an Assistant Professor in DIT University, Dehradun, in the Department of Computer Science and Engineering. She completed her MTech in Computer Science and Engineering (CSE-IS) and her BTech in Computer Science and Engineering. She is a member of Institute of Electrical and Electronics Engineers (IEEE), Computer Science Society (CSI), Association for Computing Machinery (ACM), Association for Computing Machinery - Women (ACM-W), and the Internet of Things. Her areas of research lie in machine learning, big data, biomedical, and privacy/security, in which she has supervised many student research projects. She has published more than 10 book chapters, including many in the IEEE/Springer Conferences. She has been a technical program committee member for ICIC (2018), ICETMSD (2018), ICACE (2018), WECON (2018), ICAST (2018), and RTESD (2018). She has given a talk at IIT Kharagpur in the event of Vishleshan 2019. She is a reviewer for the International Journal of Handheld Computing Research (IJHCR) and is also an editor of books published by Springer, de Gruyter. Elsevier, and many more.

Sandeep Kautish is a seasoned academician, with over 14 years of experience. He holds a PhD in Computer Science with specialization in social network analytics and sentiment analysis. Dr Kautish has over 55 peer-reviewed publications. His research interests include machine learning, data mining, and network security. One postgraduate student working under his supervision has been awarded the PhD from Bharathiar University Coimbatore. Dr Kautish is a member of the editorial boards of Inderscience, IGI Global, and the Australasian Journal of Information Systems. He has successfully organized 12 national/international conferences at Jaipur, Dehradun, and Bathinda. Dr Kautish has been nominated as a Technical Program Committee member of international conferences at Bangkok, Paris, Singapore and others, and is an active member of CSI, IEEE, CSTA, SCIEI, IACSIT, IABE, IAENG, and WASET and many more.

Dr Rajeev Tiwari is an Associate Professor in School of Computer Science (SCS) in UPES, Dehradun (India). He is a Senior IEEE Member. He was awarded his PhD in Computer Science and Engineering (CSE) from Thapar University, Patiala (Punjab). He has more than 13 years of research and teaching experience. Dr Tiwari's broad areas of research are cloud computing, MANET, VANET, QoS in wireless networks, cache invalidation techniques, Internet of Things (IoT), big data analytics, and machine learning, and is involved in funded research projects in the fields of IoT, cloud, and wireless networks. Dr Tiwari is also an expert in simulation and design of scenarios on NS2, SUMO and MOVE. He is an active member of IEEE, ACM, IEI, IACSIT, and IAENG. He has published more than thirty papers in international journals, as well as being an inventor on patents. He has chaired many sessions in international conferences. Dr Tiwari is a lead reviewer in SCI-indexed journals published by Springer, IEEE, and ScienceDirect. He has authored eight book chapters in books by international publishers.

List of Contributors

Raghav Ankur
Principal Educator
KPIT Technologies Ltd.
Bangalore, Karnataka, India

Pranav Arora
Department of Instrumentation and
    Control Technology
Netaji Subhas University of Technology
New Delhi, India

Seema Baghla
Assistant Professor (Computer Engg.)
Yadavindra College of Engineering
Punjabi University, Guru Kashi Campus
Talwandi Sabo, Bathinda, Punjab, India

Manu Bala
Student MTech. (Computer Engg.)
Yadavindra College of Engineering
Punjabi University, Guru Kashi Campus
Talwandi Sabo, Bathinda, Punjab, India

Gaurav Bathla
Chandigarh University
Gharuan, Mohali, Punjab, India

Asmita Singh Bisen
Assistant Professor, IIMT University
Greater Noida, Uttar Pradesh, India

Vinay Chowdary
Assistant Professor
University of Petroleum & Energy Studies
Dehradun, Uttrakhand, India

Hemant Kumar Gianey
Department of Computer Science &
    Information Engineering
National Chen Kung University
Tainan, Taiwan

Akhil Gupta
Assistant Professor
Electrical Engineering Department
I. K. Gujral Punjab Technical University
    Batala Campus
District Gurdaspur, Punjab, India

Gaurav Gupta
Assistant Professor (Computer Engg.)
Computer Science & Engineering
    Department
Punjabi University, Patiala, Punjab,
    India

Neelakshi Gupta
Chandigarh University
Mohali, Punjab, India

Sun-Yuan Hsieh
Department of Computer Science &
    Information Engineering
National Chen Kung University
Tainan, Taiwan

Vivek Kaundal
Specialist
Global Engineering Academy
L&T technology Services
Mysore, Karnataka, India

Gagandeep Kaur
Associate Professor
Electrical Engineering Department
I. K. Gujral Punjab Technical
    University
Main Campus, Kapurthala, Punjab,
    India

Hramanjot Kaur
Chandigarh University
Gharuan, Mohali, Punjab, India

Jaspreet Kaur
Assistant Professor
Electronics and Communication
    Engineering Department
Beant College of Engineering &
    Technology
Gurdaspur, Punjab, India.

Rohit Kumar
Govt. P.G. College
Naraingarh, Ambala, India

Nitin Mittal
Department of ECE
Chandigarh University
Mohali, Punjab, India

Akhil Nigam
Assistant Professor
Electrical Engineering Department (UIE)
Chandigarh University
Mohali, Punjab, India

Bhavesh Praveen
School of Computer Science and
    Engineering
Vellore Institute of Technology
Vellore, India

Ashish Sharma
Assistant Professor
Electrical Engineering Department
I.K. Gujral Punjab Technical University
Kapurthala, Punjab, India

Deepak Kumar Sharma
Department of Information Technology
Netaji Subhas University of Technology
New Delhi, India

Kamal Kant Sharma
Assistant Professor
Electrical Engineering Department (UIE)
Chandigarh University
Mohali, Punjab, India

Tripti Sharma
Department of ECE
Chandigarh University
Mohali, Punjab, India

P. Singh
Department of ECE
Chandigarh University
Mohali, Punjab, India

Ratanjot Singh
School of Computer Science and
    Engineering
Vellore Institute of Technology
Vellore, India

Swarnalatha P
Associate Professor
School of Computer Science and
    Engineering
Vellore Institute of Technology
Vellore, India

Sunny Vig
Assistant Professor
Electrical Engineering Department (UIE)
Chandigarh University
Mohali, Punjab, India

Tracy Austina Zacreas
Senior Principal Educator
KPIT Technologies Ltd.
Pune, Maharashtra, India

1 A Review on Localization in Wireless Sensor Networks for Static and Mobile Applications

P. Singh and Nitin Mittal

CONTENTS

1.1   INTRODUCTION

Wireless Sensor Networks (WSNs) nowadays are treated as an emerging technology, used for various applications like investigation of natural resources, tracking of static or dynamic targets, and in areas which it is not easy to access. A WSN consists of different types of sensors, which may be homogenous or heterogeneous [1]. The main challenges faced in WSNs, which degrade the performance, are computational, battery lifetime, security, and localization. The localization process is used in order to assign the coordinates to unknown target nodes deployed in the sensor field. Localization techniques can be used in WSNs for different applications, such as tracking of targets and location tracking of target nodes, etc. Many researchers have presented a variety of localization algorithms for improving important parameters, namely accuracy and efficiency. These techniques are mainly classified as either range-based or range-free localizations. Techniques, such as received signal strength indicator (RSSI) [2, 3], time of arrival (TOA) [4, 5], time difference of arrival (TDOA) [6], angle of arrival (AOA) [7, 8], are classified as range-based techniques. The range-based localization techniques are used for calculating the position of the node with range information (based on angle or distance). A huge cost is involved in implementing range-based methods but these methods are more effective at localizing the node effectively and for guaranteeing accurate node localization, as compared with range-free techniques. Some range-free techniques are classified as Centroid Method [9], DV-Hop [10], approximate point in triangulation (APIT) [11] and multidimensonal scaling (MDS) [12]. The benefit of using range-free methods is their ease of operation and low overheads.

In WSN, deployment is not always static. It may also be dynamic, but there are a few problems that need to be overcome, like the maintenance of connectivity, scope, and utilization of energy. The current trend in today's WSNs put mobility in a positive light. Localization is the main requirement as well as the biggest challenge for dynamic WSNs. The accurate positions of the nodes placed in the sensor field must be known in order to identify the most-efficient route. The sensor nodes may also move from one point to another during their run-time, in the case of dynamic WSNs, but the position of sensor nodes is not going to vary from its original position in the case of static WSNs. Therefore, localization of sensor nodes in dynamic environments is of prime concern.

The rest of this chapter is organized as follows: Section 1.2 represents related work, Section 1.3 represents classification of localization algorithms, and Section 1.4 represents the challenges faced in localization. Section 1.5 represents determination of 2-D and 3-D coordinates, whereas Section 1.6 represents applications, and Section 1.7 represents the techniques used to improve localization. Finally, the conclusions of current research in this topic are represented in Section 1.9.

1.2 RELATED WORK

To localize the sensor nodes, to enhance their various parameters in order to increase their lifetime and accuracy, localization plays an important role, becoming an essential requirement for multiple applications in the field of WSNs. The literature provides detailed explanations on various localization techniques available to perform this task. Mainly, localization algorithms consist of two stages: the first stage is used for measuring the distance and the second stage is used for solving the computations.

1.2.1 MEASUREMENT STAGE

In this stage, the distance measured between different nodes is considered to be an important parameter, including angle measurement, and their own connectivity between them is considered. These techniques are classified into five main categories described as follows:

1. Strength of the received signal.
2. Arrival of the signal at a particular time or the difference in their arrival times.
3. Angle of arrival.
4. Proximity, based on the network connectivity.
5. Picture/scene analysis.

Each category will be described in some detail.

Received Signal Strength Indication (RSSI). This is used basically to observe the received incoming signal. On the arrival of the received signal, the task of RSSI is to calculate the distance on the basis of the incoming signal. Distance calculations are performed, using the received signal strength of the incoming signal [13, 14].

$$P_{\text{receive}} = c.\frac{P_{\text{trans}}}{d^{\alpha}} \tag{1.1}$$

$$d = \sqrt[\alpha]{c.\frac{P_{\text{trans}}}{P_{\text{receive}}}} \tag{1.2}$$

In Eq. (1.1) and (1.2), P_{trans} and $P_{receive}$ represent the transmitted power and the received power, respectively, and α represents the path loss coefficient. In this equation, d represents the distance between transmitter (Tx) and receiver (Rx) and is calculated using the received signal power. Only minimal hardware is needed to calculate the received signal strength, which is the main advantage of this method. But its limitation is that the measured values are going to change in the case of mobility, environmental conditions, path loss, or fading.

- ToA: It finds the distance and time at which the incoming signal is received, as the speed of propagation information is already available. Both the transmitter and the receiver end are synchronized and the start time of transmission is known. The parameter c (speed of the light) is known and distance calculation is done, based on the time of the incoming signal's arrival at the receiver end [15, 16]. In this method, synchronization must be required between the transmitter and the receiver clocks for accuracy, which leads to additional hardware requirements at both units, and increases complexity. Fig. 1.1 illustrates the concept of ToA.
- TDoA: In this scheme, the medium used for transmission gives a different speed. As shown in Fig. 1.2, the radio and the ultrasound signals are transmitted from the transmitter end. The arrival of any signal at the receiver end is used to measure the arrival of the other signal. The distance is calculated based on the arrival time of these two transmitted signals. This method provides accuracy under line of sight (LOS) conditions but, if certain disturbances occur in the environmental conditions, it leads to the failure of LOS conditions. Also, with the change in environmental conditions,

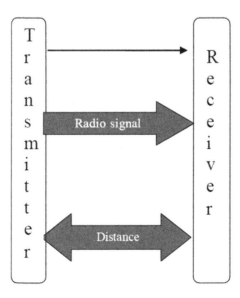

FIGURE 1.1   Arrival time of a signal

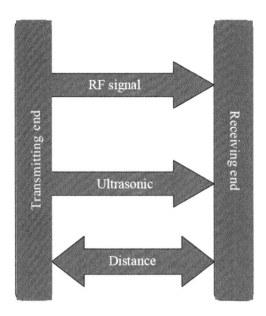

FIGURE 1.2 Difference in the time of arrival

these arrival times are going to vary from their actual values which leads to incorrect distance measurements.

- Angle of Arrival (AoA): The distance calculation between the anchor and the target nodes is also calculated using the Angle of Arrival method. In this method, an angle is made between the two lines, where the first line connects the transmitting and receiving ends and second line is between the receiver and the other direction, taken as a reference, as shown in Fig. 1.3. The distance measured through this technique is more accurate than the distance calculated using the RSSI method mentioned above [17].

- Proximity: This is the simplest and cheapest method available for calculating the distance between the nodes because it measures distances only between those nodes which are inter-connected and within range. The hardware configuration required is simplest under this scheme [18].

- Picture Analysis: This technique behaves in a different manner from the above-mentioned techniques. In this, the distance calculations are carried out using a picture or on the basis of scene analysis. In this technique, additional hardware is required which leads to complexity and is the main drawback of this technique.

1.2.2 COMPUTATIONAL STAGE

In the second, computational stage, the estimations done on determining the distance and angles in the previous stage are collated for calculating the positions of

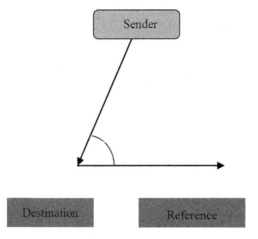

FIGURE 1.3   Signal received at a particular angle

the unknown nodes. These methods, which work on computational analysis, are discussed as follows.

- Trilateration: In this scheme the estimated coordinates of the target node is determined with the help of three anchor nodes, and the location of the tracking nodes is calculated [19]. As shown in Fig. 1.4, the determination of the target nodes coordinates is obtained from the coincidence of consecutive circles. In the case of determining the 2-D coordinates, at least three anchor nodes are required.

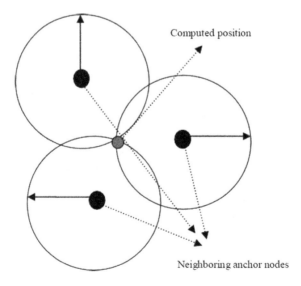

FIGURE 1.4   Trilateration

- Triangulation: Geometrically, the triangulation technique is used to obtain the information of 2-D coordinates on the basis of angles calculated between the nodes and the reference points. By using mathematical sine and cosine rules, the position of the target nodes can be calculated [20]. Fig. 1.5 shows this technique.
- Multilateration: In the trilateration approach, the calculated distance is not perfect because the joining of the three circles does not correspond to a single point. In order to cope with this limitation, at least three anchor nodes are required, a process termed multilateration [21]. The results of this technique are much more efficient than those from trilateration. Fig. 1.6 represents this concept.

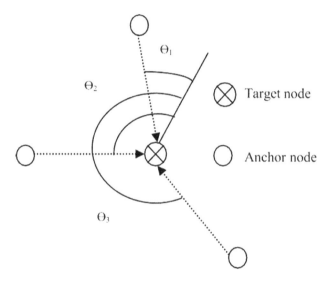

FIGURE 1.5   Concept of triangulation

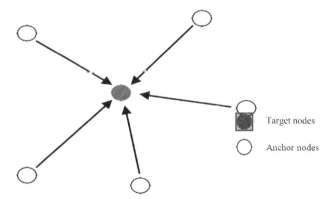

FIGURE 1.6   Multilateration concept

Three anchor nodes are assumed with their known position, i.e., (x_i, y_i) for $i = 1,2,3$ and the unknown position of the target nodes (x_t, y_t) is calculated by using the equation as given:

$$d_{i,t} = \sqrt{(x_t - x_i)^2 + (y_t - y_i)^2} \tag{1.3}$$

where $d_{i,t}$ represents the distance between the target nodes and the anchor nodes.

1.3 CLASSIFICATION OF LOCALIZATION ALGORITHMS AVAILABLE IN WSNS

The localization concept in sensor networks is gaining importance in almost all real-world applications of WSNs. The survey of these types of algorithms provides detailed explanations of the different techniques available for localizing the nodes referred to as anchor-based, anchor-free, range-free and range-based nodes, etc. Every node connected within the network transmits a beacon signal, that will be analyzed by the receiver at the receiver section in order to measure the range, which will help in counting the number of hops. Fig. 1.7 classifies the various available localization algorithms.

1.3.1 LOCALIZATION BASED ON HOP COUNTS

In single-hop localizations, the unknown nodes, called target nodes, are localized in that they are only one hop away from the known positions of the anchor nodes,

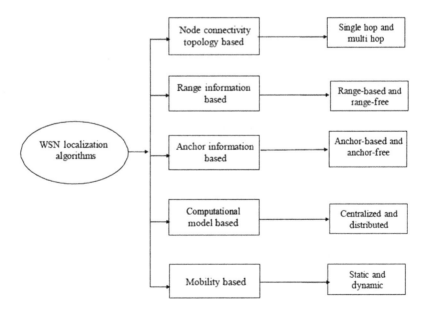

FIGURE 1.7 Classification of localization

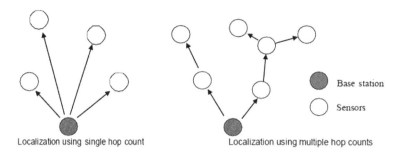

FIGURE 1.8 Localization using hop counts

whereas, in the case of multiple hop schemes, the unknown nodes and the known anchor nodes are more than one hop count apart; multiple hop counts are those which connect the two nodes. The methods used for single-hop counts are Angle of Arrival, Time of Arrival, Time Difference of Arrival, and Received Signal Strength Indicator, whereas the multi-hop method covers Distance Vector Hopping scheme, Approximation of Points Residing in the Triangles, and Multidimensional Scaling. Fig. 1.8 represents this concept. In these methods, additional numbers of anchor nodes are required, which is the major drawback [22–25].

A matrix, symmetric in nature, is applied as an input to the multi-hop WSN localization method. In this, hop counts on the basis of distance between the nodes are calculated and used. The techniques used under this scheme are referred to as range-based and range-free localizations.

1.3.2 LOCALIZATION BASED ON RANGE-BASED AND RANGE-FREE TECHNIQUES

Localization algorithms are divided into two categories, namely range-based and range-free techniques. In range-based localization techniques, the anchor node's range information is required, but, in the second method, information relevant to range is not required. By using some specialized hardware, the distances between the nodes is calculated precisely. The main issue arising from this technique is accuracy degradation taking place in case of mobility scenario and of various noises taking place in the environment.

A network formation in the case of the range-free technique is that in which the hop count is directly proportional to the distance between them. In case of range-free techniques, the coordinates are calculated on the basis of radio connectivity information among their available neighboring nodes. This technique is more efficient, in terms of simplicity and cost-effectiveness, than the range-based localization techniques. Several approaches related to range-free techniques are given below [26, 27].

1. Based on the Centroid method: This method was developed by Blusu et al. [28]. In this algorithm, at least three anchor nodes are required in order to locate the position of the unknown node, which are in its neighborhood position. Eq. (1.4) represents the method of calculating the centroid, where

$x_1 + x_2 + \cdots x_N$ and $y_1 + y_2 + \cdots y_N$ represent the x and y coordinates of the anchor nodes and x_{est}, y_{est} represents the estimated positions of the unknown node, referred to as the target node.

$$x_{\text{est}}, y_{\text{est}} = \frac{x_1 + x_2 + \cdots x_N}{N} + \frac{y_1 + y_2 + \cdots y_N}{N} \tag{1.4}$$

2. DV-Hop method: In the DV-Hop method, the hop count between two anchor nodes is calculated and, from this information obtained, the length of each hop is calculated. The information gathered by every anchor node is floated or shared into the sensing field area. This information will help locate the target nodes by estimating the multi-hop range [26, 29, 30].

3. Approximating Point in Triangular method (APIT): In this technique, the area under investigation is segregated into triangular regions. A node lying near or far from the triangular region resides in the corresponding triangle by narrowing down its area [11].

4. Multi-dimensional scaling (MDS): In this method, again on the basis of distance, all nodes will maintain a distance matrix. After estimating the distance matrix, Dijkstra's algorithm for finding the shortest path is used to design a complete matrix. When the observation accuracy of a sensor becomes worse, this technique is not efficient during that time, then a hybrid technique, in which this method is used in combination with trilateration, can provide the efficient results [31].

1.3.3   LOCALIZATION BASED ON ANCHOR-BASED AND ANCHOR-FREE TECHNIQUES

These are the two techniques available on which the location of unknown nodes is relied upon. In the first (anchor-based) method, the information on the anchor node is required in order to calculate the coordinates of the unknown node; in the case of anchor-free techniques, no such information about the anchor is required. In the deployment stage the position of a few nodes, known as anchor nodes, are known, as they include a GPS feature, whereas, in the case of the anchor-free method, localization is achieved, using the information of the relative coordinates [32, 33]. The main motive behind the use of anchor-based method is the calculation of the distance from the unknown nodes to the known nodes; after the calculation on this basis is carried out between them, then, by using localization algorithms, the unknown node determines its coordinates in space, as, for determining 2-D coordinates, at least three anchor nodes are required and, for determining 3-D coordinates, four anchor nodes at least are required.

In cases where no anchor node is available, there are two important steps to be followed for determining or localizing the unknown nodes.

By the use of different localization methods available, every node in the sensing field computes the distance between them and their neighboring nodes.

Then, on the basis of distance, every node in the sensing field determines its coordinates itself.

1.3.4 CENTRALIZED AND DISTRIBUTED LOCALIZATION

A central processor is responsible for computing all the computations by the centralized method, but, in the latter method, the computations are performed using the inner sensor measurements. The main advantage of using the centralized method is that each node does not need to perform the computations. The major drawback of using this technique is that all the nodes send their data back to the base station. Simulated annealing and RSSI-based localization come under the classification of centralized localization algorithms. These algorithms are better in terms of localization accuracy because complete information about the connectivity and distance is available between the deployed sensor nodes and their neighbors [34].

In the case of distributed localization algorithms, the sensors deployed in the field compute the required information in terms of either connectivity or distance on an individual basis. In this method, each node communicates with them or with their neighbors, in order to determine their own position in the network. Beacon-based, coordinate system-based and hybrid localizations are some of the examples of distributed localization algorithms. In this method, many iterations are required to achieve stability, resulting in the technique being a little slow, which leads to a drawback of this method [35].

1.3.5 STATIC AND DYNAMIC LOCALIZATION

In the static algorithms, once the nodes are deployed in the region of interest, they will not move; the dynamic method nodes deployed in the sensing field have some mobility and will move accordingly [36–38]. In static algorithms, the coordinates of unknown nodes are determined during the set-up of static WSNs. But in the case of the dynamic method, a continuous tracking of sensor nodes is required as they are moving and changing their coordinate values. A little extra time is required to determine the coordinates of the moving nodes; for tracking the position of moving nodes, a fast localization feature is required, using this procedure. In the case of static algorithms, the convergence rate is fast. There are many range-based techniques available, like approximate point in triangulation and lateration, multilateration, and the modified centroid method, which are some of the static localization techniques.

It is very difficult to track the location of moving target nodes because it has to be determined periodically. There are various techniques available in the literature, of which the Kalman filter is one which works well in noisy environments. Road navigation in the absence of GPS features is a property of Kalman Filters [39, 40]. In order to predict the system's future states, the Kalman filter is one of the useful techniques.

These issues are described in the following section.

1.4 LOCALIZATION CHALLENGES

In many real-time applications, the localization concept is becoming an important and essential requirement in the field of WSN. The available literature presents many challenges to be faced, while using the concepts of localization [1]. As represented in Fig. 1.9, there are some issues that need to be taken care of in order to improve the localization process of the WSNs.

1.4.1 ENERGY OF THE NODE

The sensor nodes deployed in the sensing field in the area of WSNs have non-replaceable and limited-energy battery units for performing the operations, such as sensing and reporting. The whole system becomes worse if special care is not taken with regard to battery units. A lot of research has been conducted in order to improve the energy of the node [41].

1.4.2 NODE MOBILITY

In order to maintain the node's connectivity in the mobility scenario is quite a challenging task in localization. In the case of static WSNs, the node, once estimated, is not going to change its position as it is fixed, but, in the case of dynamic WSNs, the nodes periodically shift from one position to another position, and the deployed sensor node has to determine and estimate the position periodically [36, 38]. So, localization is becoming a major source of interest amongst researchers.

1.4.3 TRANSMISSION RANGE OF NODE

In most of the localization techniques, a beacon signal is required for determining the locations of unknown nodes. Beacon nodes are equipped with a GPS feature which helps these nodes to be placed at a position with known coordinates, and, with this feature, one can obtain a proper connectivity in between the beacon and the

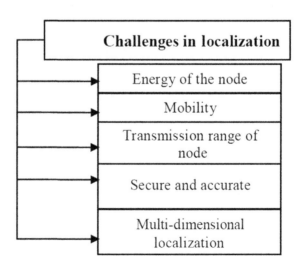

FIGURE 1.9 Localization challenges

target nodes. So, the transmission range of a node plays a key role in estimating the locations of unknown nodes.

1.4.4   LOCALIZATION SECURITY

As in WSNs, most of the times this set-up is installed in unfriendly locations. Some of the problems that occur in localization are overshadowing and distance away from attack [42]. So, researchers are showing interest in improving the security of localization.

1.4.5   LOCALIZATION ACCURACY

Accuracy is determined by calculating the difference between the actual and the estimated position of the sensor node. It is a difficult task to obtain the accurate location of the sensor node by applying localization algorithms. So, by using available or new localization algorithms, it is possible to obtain optimum results [43, 44].

1.4.6   MULTI-DIMENSIONAL LOCALIZATION

Basically, in order to obtain the coordinates of unknown nodes (target nodes), localization algorithms are used. Some of the applications require knowledge of 2D and 3D coordinates [36].

1.5   DETERMINATION OF THE 2-D AND 3-D COORDINATES OF THE SENSOR NODE

As many advances and improvements in the field of wireless communication have taken place in recent years, this has encouraged the use of WSNs in many real-world applications. Localization of sensor nodes becomes necessary in almost every application in WSNs. There are various localization algorithms available in the literature, which will determine the location of target nodes placed in the sensing field, which, in turn, are possible with or without the use of anchor nodes.

1.5.1   DETERMINATION OF 2-D COORDINATES

Bulusu et al. [28], in their research, investigated localization in outdoor conditions without using GPS features. They used the centroid method for calculating the coordinates of the unknown nodes. Lee et al. [45] represented their work on localization by using very few anchors in the field. Their work presented and claimed a higher accuracy in estimating the distance by finding the shortest path. Awad et al. [46] presented two ways for estimating the distance; the first way was to use the available statistical methods and the other one was to use the concepts of neural networks for computing the distance. Their conclusions were based on the parameters which degrade the quality of distance measurements, such as transmitted power, RF frequencies, mobility of nodes, and localization algorithms. Savvides et al. [47], in their work, discussed the use of the multilateration technique. The authors presented a distributive approach in that sensor nodes were collectively going to solve an optimization problem, which a single sensor could not solve. Three state problems are discussed in this paper. In state-1, a combined approach was taken, the second state represents the computation of starting estimates, and, in the final state, again, reframing of positions was carried out. Savvides et al. [48] presented an approach to localizing sensors in a temporary network that allows the deployed sensors in the field to find their locations using distributive methods. Although location estimates are more accurate by using centralized methods, by using a distributive approach, the system is more robust and will result in an effective distribution of power consumption in the network. Sumathi and Srinivasan [49] used the least squares method for estimating the locations of static target nodes by the use of single anchor nodes, on the basis of RSSI measurements.

Guo et al. [50] presented a different approach in which there is no relation or mapping of estimating distances based on RSSI values, using the method named as the Perpendicular Intersection method. Using this geometric relationship, the positions of the nodes are computed. Kim and Lee [51] presented a localization technique that is range based and in which the movement of the mobile anchor node is based on some strategy, known as MBAL (Mobile Beacon-Assisted Localization). Their scheme provided the best selection of the path, with fewer complexities.

In their conclusions, Karim et al. [52] discussed the range-free techniques that are efficient in terms of energy with the use of fewer anchor nodes. They claimed that their technique is less complex and more accurate by using very few anchor nodes. Li et al. [53] proposed a method that provides efficient and robust localization in real-time scenarios. The techniques used are named the Breadth-First (BRF) and Backtracking Greedy (BTG) algorithms. Chen et al. [54] presented an approach that is classified under the range-free methods. Their work stated that their scheme is simple and practical, in which a comparison among the nodes is based on RSSI values and this consumes much less energy and shows great accuracy, while using the concepts of mobile anchors. Khelifi et al. [55] proposed two different localization algorithms, namely time-driven and event-driven for mobile WSNs. Stone and Camp [56] discussed the localization algorithms that use the information of anchors, and computed the performance and accuracy of localization relative to anchor mobility and target nodes. There is much more research which we have not mentioned, and conclusions are available in the literature [57–63].

1.5.2 DETERMINATION OF 3-D COORDINATES

Shi et al. [64] presented Ultra-Wide Band, a Time of Arrival method determining the 3-D coordinates required for localization. They discussed that distance calculation, using their technique between the anchor node and the unknown node, can be more precise than other techniques. Wang et al. [65] presented a scheme based on the Distance Vector hopping algorithm, that can localize 3-D coordinates of a sensor node more precisely and effectively, but its complexity and huge deployment cost are its major limitations. Xu et al. [66] discussed a hybrid method by using Distance Vector hopping combined with the Newton method for optimization. The authors considered two parameters, one of which is coverage and the other accuracy, for their proposed algorithm. Li et al. [67] developed a model for determining 3-D coordinates in a WSN environment on the basis of the RSSI model. They proposed a model for obtaining relevance between the Degree of Irregularity and variation in the ranges of the transmitted signal. Ahmad et al. [68] developed an algorithm for the determination of 3-D coordinates, using a parametric method. In their algorithm, very few anchor nodes are available for localization and will provide better results because the network is shrinking towards a common point known as the central point. Zhang et al. [69] proposed a localization algorithm that depended upon an expensive beacon signal. MDS-MAP, DVHOP, and the Centroid method are methods modified from 2-D coordinates to 3-D coordinates, which are available in the literature [71–75]. In obtaining the 3-D coordinates in applications deployed in an underwater network [76], Cheng et al. [77], and Zhou et al. [78] surveyed this application. In these methods, localization is achieved using information based on connectivity and the number of anchors. Tomic et al. [79] presented a hybrid method in developing 3-D WSNs. In this scheme, an estimate on the basis of least squares criterion is used. Chan et al. [80] proposed a method for solving a problem based on transmission power as some 3-D localization methods require specific directional information, by implementing a directional antenna on anchor nodes [81, 82].

1.6 APPLICATIONS

Navigational services are now being used by everyone everywhere in the world, but accurate positioning of mobile devices is becoming a current research topic.

1.6.1 SERVICES BASED ON LOCATION

Through mobile networks and/or the internet, the information about the location is provided to end users. Applications based on navigational services can provide the connectivity to link the position of the mobile user, with geographical locations tracked by mobile users. These services are essential in both indoor and outdoor environments. These types of services are required for guiding the users to a particular area of interest. Also, they are used in a public place, e.g., in railway stations or bus stations, to direct the passengers to a particular desired place.

1.6.2 HEALTH AND SMART LIVING APPLICATIONS

Localization in the indoor environments plays a great role for these types of applications. These types of applications are helpful for determining and monitoring the health of elderly people [83].

1.6.3 ROBOTICS

Robotics is one of the important applications in the field of localization. There is much research going on for the development of these types of applications. The cooperation between the installed robots in the field is required for certain applications like surveillance and for the tracking of other unknown areas [84].

1.6.4 CELLULAR TECHNOLOGY

Many challenges and issues about the accurate determination of location are determined by cellular technology [85], as the growth in the mobile generation leads to an improvement in the accuracy.

1.7 TECHNIQUES USED TO IMPROVE LOCALIZATION

1.7.1 FEATURES OF OPTIMIZATION TECHNIQUES

The features of optimization techniques are listed below.

1.7.1.1 Speed

In real-time environments, the convergence rate of the localization algorithm used should be fast enough.

1.7.1.2 Adaptability

The system should adapt to the environment and report accordingly, in case the node fails.

1.7.1.3 Self-Organizing Capability

The system will have self-organizing capabilities, meaning that, in cases of mobility, it will re-organize the network itself. Some pre-defined rules and instructions are posed on the network, so that system will adapt the changes accordingly.

1.7.2 GENETIC ALGORITHMS (GA)

This is a technique based on search and optimization and is used for finding the estimated results. It begins its search with random solutions and these solutions are assigned a fitness function that is relative to their objective function. Then, a set of new populations is formed by using three genetic operators, named as reproduction, cross-over and mutation [86]. An iterative operation in GA takes place using all three operators until a terminated criterion is not reached. For decades, GA has been used in a wide variety of applications, because of its simplicity. The basic flow of genetic algorithms is given in Fig. 1.10.

1.7.3 PARTICLE SWARM OPTIMIZATION (PSO)

A technique developed by Kennedy and Eberhart is known as PSO [87], based on the behavior of birds. It is an efficient algorithm and its implementation stage is easier. A random number of particles is deployed in the search space and the objective function is calculated accordingly. Then, the movement is applied to the particles deployed in the search space [88]. A particle moving in the search space collects 'pbest' and 'gbest' positions in the space. The idea is illustrated in Fig. 1.11.

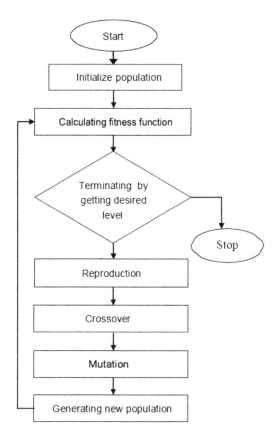

FIGURE 1.10   Genetic algorithm

1.7.4   BIOGEOGRAPHY BASED OPTIMIZATION (BBO)

In BBO, the term HIS (Habitat Suitability Index) represents the fitness function. The higher the value of HIS, the better the place is for the species to live, whereas lower HIS values indicate an inappropriate place for the species to live. The basic idea about the BBO algorithm is shown in Fig. 1.12.

1.7.5   FIREFLY ALGORITHM

The firefly algorithm was proposed by Yang [89]. The behavior of fireflies is used in this algorithm and the rules followed by the fireflies are as follows:

   All fireflies are unisex as they move from one place to another notwithstanding of sex [90].
   The parameter which attracts the fireflies towards each other is attractiveness, which is directly proportional to the glowing nature of the fireflies, and, as they move a certain distance apart, their brightness is reduced. So, fireflies will not follow each other in that particular case. If there is no brighter

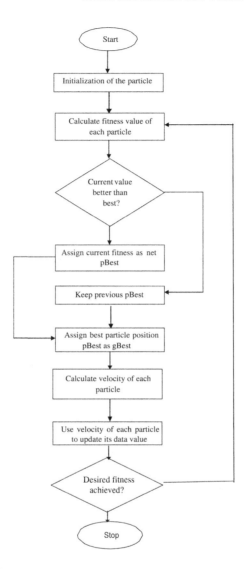

FIGURE 1.11 PSO

firefly found, then this event is random in nature. The fitness function in
this case is represented by the glowing nature of fireflies. Then, according
to these rules, the firefly algorithm is represented in Fig. 1.13.

According to the Free Lunch theory, no single algorithm is best-suited to each optimization problem. There are many more types of optimization algorithms that are
reported in the literature, which can be applied to localization problems to check
their performance.

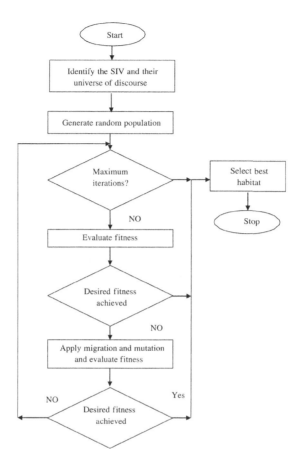

FIGURE 1.12 BBO

1.8 CRITERIA FOR EVALUATION AND PERFORMANCE PARAMETERS

In WSNs, multiple errors occur, such as errors due to range problems, errors due to non-availability of GPS signals and sometimes localization algorithms, which are used for certain applications, degrade the accuracy of the system. Range error arises due to incorrect measurements carried out on the basis of distances. Similarly, errors in anchor position lead to a GPS error. There are certain parameters on which the performance of the algorithms depends on accuracy in terms of location, its cost and coverage. The evaluation procedure is described in Fig. 1.14.

- Accuracy in Location: In this, the difference between the node's original position and the estimated position is calculated using any localization algorithm and the difference between the two leads to an error. By using

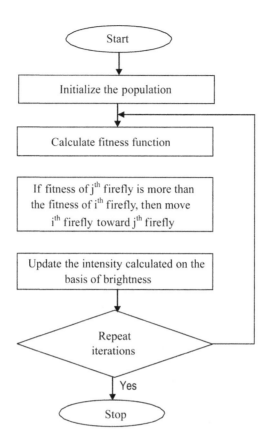

FIGURE 1.13 Firefly algorithm

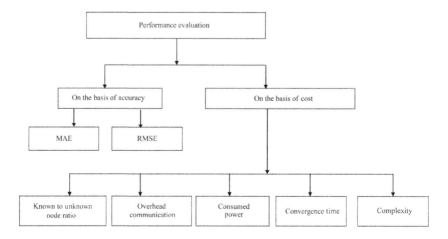

FIGURE 1.14 Procedure for evaluating the performance

this information, the determination of the accuracy parameter is calculated; the smaller the error, the greater will be the accuracy, and *vice versa*.

- Flexibility to Error and Noise: The localization algorithm chosen for the determination of location should be flexible enough to combat errors or noises originating from the input side.
- Coverage: The coverage parameter depends upon a few conditions, such as the number of anchor nodes deployed in the sensing field. The larger the number, the better the coverage.
- Cost: In this, the cost parameter is evaluated on the basis of power consumed and the time taken by the algorithms to localize the nodes, so that communication between the nodes can be initiated.

1.8.1 Parameter Calculations on the Basis of Accuracy

The accuracy term is used to match the positions of the actual and the estimated target nodes. The difference between the two positions leads to errors and these errors are named the Mean Absolute Error and the Root Mean Square Error.

Mean Absolute Error (MAE): MAE is basically calculated with respect to the continuous variables and it is an important parameter for finding out the accuracy of a localization algorithm used in a specified application. The equation for MAE is Eq. (1.5), where (x_t, y_t, z_t) is the current position, (x_e, y_e, z_e) is the calculated position, and N_t represents the total number of sensor nodes deployed.

$$\text{Absolute Error} = \frac{\sum_{t=1}^{Ni} \sqrt{(x_t - x_e)^2 + (y_t - y_e)^2 + (z_t - z_e)^2}}{N_t} \tag{1.5}$$

Root Mean Square Error (RMSE): This parameter also represents a measure of accuracy and is given by Eq. (1.6),

$$\text{RMSE} = \sqrt{\frac{\sum_{t=1}^{Ni} (x_t - x_e)^2 + (y_t - y_e)^2 + (z_t - z_e)^2}{N_t}} \tag{1.6}$$

1.8.2 Evaluation on the Basis of Cost Metrics

In almost every application, cost factor determination plays an important role. In the context of localization, the parameters which contribute to the cost are power consumed during the set-up stage, how many anchor nodes are required for this process, and the total time required to localize all the nodes in the network. If lifetime enhancement of the network is important at the same time, cost management also plays an equal role. There is a trade-off between these two. The ratio of known positions of the anchor nodes to the unknown positions, power consumed, and time taken by the localization algorithm to localize all the nodes play important roles in determining cost metrics.

- Anchor to Target Ratio: In terms of cost metrics, anchor nodes are the ones which are deployed in the sensing field with GPS features enabled in them. In order to save the cost, we need to install only a few anchor nodes in the field as they are expensive, and, in order to localize the target nodes, this terminology is used. It is defined as the number of anchor nodes required to localize the target nodes.
- Overhead during Communication: As the number of sensor nodes deployed in the sensor field increases, the communication overhead also increases to a greater extent. The overhead can be calculated by finding out the total number of packets sent.
- Convergence Time: The time taken by the localization algorithm to collect all the information regarding localizing all the nodes present in the network represents the convergence time. As the network size increases, this parameter is affected.
- Algorithmic Complexity: Algorithmic complexity is always defined with some standard notions (O), where the higher the order, like $O(n^3)$ and $O(n^2)$, the longer time it will take to converge, with this parameter representing the complexity.
- Power Consumed: This parameter is important in terms of cost as it calculates the power consumed in a localization process.

1.9   CONCLUSIONS

In this chapter, the two different scenarios, the Static and Dynamic issues of WSNs, are discussed. There are two main stages on which estimation of a sensor node's location depends, i.e., the measurement stage and the computational stage. Furthermore, they are classified on the basis of range based or range free, anchor based or anchor free, and the mobility scenario has already been discussed. There are certain challenges faced in the localization process. The major challenge is to localize the sensor node's location in 2-D and 3-D scenarios. Many optimization techniques have been used to estimate the accurate locations of the sensor nodes. Still, there are open questions in this research area, like the localization in mobility-based scenarios and the use of smaller numbers of anchor nodes to save costs. Therefore, one can introduce different optimization techniques in the future to solve the various issues arising in the localization process.

BIBLIOGRAPHY

1. Karl, H, Willig, A (2007) *Protocols and Architectures for Wireless Sensor Networks.* John Wiley & Sons.
2. Cheng, L, Wu, C D Zhang Indoor robot localization based on wireless sensor networks. *IEEE Trans. Consum. Electron.* 57(3), 1099–1104.
3. Barsocchi, P, Lenzi, S, Chessa, S, Giunta G (2009) A novel approach to indoor RSSI localization by automatic calibration of the wireless propagation model. In: *IEEE 69th Vehicular Technology Conference*, pp. 1–5, VTC Spring.

4. Chan, Y T, Tsui, W Y, So, H C Ching (2006) Time-of-arrival based localization under NLOS conditions. *IEEE Trans. Veh. Technol.* 55(1), 17–24.
5. Xu, E Ding, Dasgupta, Z (2011) Source localization in wireless sensor networks from signal time-of-arrival measurements. *IEEE Trans. Sig. Process.* 59(6), 2887–2897.
6. Gillette, M D, Silverman, H F (2008) A linear closed-form algorithm for source localization from time-differences of arrival. *IEEE Sig. Process. Lett.* 15, 1–4.
7. Rong, P Sichitiu (2006) Angle of arrival localization for wireless sensor networks. In: *3rd Annual IEEE Communications Society on Sensor and Ad Hoc Communications and Networks, SECON'06*, vol. 1, pp. 374–382.
8. Kułakowski, P, ValesAlonso J (2010) Angle-of-arrival localization based on antenna arrays for wireless sensor networks. *Comput. Electr. Eng.* 36(6), 1181–1186.
9. Deng, B Huang, Zhang, G, Liu, L H (2008) Improved centroid localization algorithm in WSNs. In: *3rd International Conference on Intelligent System and Knowledge Engineering, ISKE 2008*, vol. 1, pp. 1260–1264.
10. Chen, K Wang *An Improved dv-Hop Localization Algorithm for Wireless Sensor Networks* Hindawi.
11. Wang, Zeng, Jin, J H (2009) Improvement on APIT localization algorithms for wireless sensor networks. In: *International Conference on Networks Security, Wireless Communications and Trusted Computing, NSWCTC'09*, vol. 1, pp. 719–723.
12. Shang, Y, Ruml W (2004) *Improved MDS-Based Localization in Twenty-Third Annual Joint Conference of the IEEE Computer and Communications Societies, Infocom*, vol. 4, pp. 2640–2651.
13. Alippi, C, Vanini, G (2006) A RSSI-based and calibrated centralized localization technique for wireless sensor networks. In: *Fourth Annual IEEE International Conference on Pervasive Computing and Communications Workshops, 2006. PerCom Workshops 2006*, p. 5.
14. Deng Yin, Z, Guo-Dong, C (2012) A union node localization algorithm based on RSSI and DV-Hop for WSNs. In: *Second International Conference on Instrumentation, Measurement, Computer, Communication and Control (IMCCC)*, pp. 1094–1098.
15. Humphrey, D, Hedley, M (2008) Super-resolution time of arrival for indoor localization. In: *IEEE International Conference on Communications*, pp. 3286–3290.
16. Cheung, K W, So, H C, Ma, W (2004) Least squares algorithms for time-of- arrival-based mobile location. *IEEE Trans. Sig. Process.* 52(4), 1121–1130.
17. Azzouzi, S, Cremer, M, Dettmar, U, Kronberger, R, Knie, T (2011) New measurement results for the localization of UHF RFID transponders using an angle of arrival (AOA) approach. In: *IEEE International Conference on RFID (RFID)*, pp. 91–97.
18. Torre, A, Rallet, A (2005) Proximity and localization. *Reg. Stud.* 39(1), 47–59.
19. Boukerche, A, Oliveira, H A, Nakamura, E F, Loureiro, A A Localization systems for wireless sensor networks. *IEEE Wirel. Commun.* 14(6), 6930–6952.
20. Tekdas, O, Isler, V (2010) Sensor placement for triangulation-based localization. *IEEE Trans. Automat. Sci. Eng.* 7(3), 681–685.
21. Zhou, Y, Li, J, Lamont, L (2012) Multilateration localization in the presence of anchor location uncertainties. In: *IEEE Global Communications Conference (GLOBECOM)*, pp. 309–314.
22. Bal, M, Liu, M, Shen, W, Ghenniwa, H (2009) Localization in cooperative wireless sensor networks: A review. In: *13th International Conference on Computer Supported Cooperative Work in Design, CSCWD*, pp. 438–443.
23. Whitehouse, K, Karlof, C, Culler, D (2007) A practical evaluation of radio signal strength for ranging-based localization. ACM SIGMOBILE *Mob. Comput. Commun. Rev.* 11(1), 41–52.

24. Franceschini, F, Galetto, M, Maisano, D, Mastrogiacomo, L (2009) A review of localization algorithms for distributed wireless sensor networks in manufacturing. *Int. J. Comput. Integr. Manufact.* 22(7), 698–716.

25. Lakafosis, V, Tentzeris, M M From single-to multihop the status of wireless localization. *IEEE Microw. Mag.* 10(7), 34–41.

26. He, T, Huang, C, Blum, B M, Stankovic, J A, Abdelzaher, T (2003) Range-free localization schemes for large scale sensor networks. In: *Proceedings of the 9th Annual International Conference on Mobile Computing and Networking*, pp. 81–95.

27. Dil, B, Dulman, S, Havinga, P (2006) Range-based localization in mobile sensor networks. *Wirel. Sensor Netw.* 0302-9743, 164–179.

28. Bulusu, N, Heidemann, J, Estrin, D (2000) GPS-less low-cost outdoor localization for very small devices. *IEEE Pers. Commun.* 7(5), 28–34.

29. Kumar, S, Lobiyal, D (2013) An advanced DV-Hop localization algorithm for wireless sensor networks. *Wirel. Pers. Commun.* 71, 1–21.

30. Tian, S, Zhang, X, Liu, P, Sun, P, Wang, X (2007) A RSSI-based DV-Hop algorithm for wireless sensor networks. In: *International Conference on Wireless Communications, Networking and Mobile Computing, WiCom*, pp. 2555–2558.

31. Stojkoska, B R, Kirandziska (2013) Improved MDS-based algorithm for nodes localization in wireless sensor networks. In: *EUROCON 2013 IEEE*, pp. 608–613.

32. Priyantha, N B, Balakrishnan, H, Demaine, E Teller (2003) Anchor-free distributed localization in sensor networks. In: *Proceedings of the 1st International Conference on Embedded Networked Sensor Systems*, pp. 340–341.

33. Ssu, K F, Ou, C H, Jiau, H C (2005) Localization with mobile anchor points in wireless sensor networks. *IEEE Trans. Veh. Technol.* 54(3), 1187–1197.

34. Zhang, Q Huang, Wang, J, Jin, J, Ye, C, Zhang, J W (2008) A new centralized localization algorithm for wireless sensor network. In: *Third International Conference on Communications and Networking in China, 2008. ChinaCom 2008*, pp. 625–629.

35. Langendoen, K, Reijers, N (2003) Distributed localization in wireless sensor networks: A quantitative comparison. *Comput. Netw.* 43(4), 499–518.

36. Singh, P, Khosla, A Kumar A Khosla 3d localization of moving target nodes using single anchor node in anisotropic wireless sensor networks. *AEU Int. J. Electron. Commun.* 82, 543–552.

37. Parulpreet, S Arun, Anil, K, Mamta, K (2017) Wireless sensor network localization and its location optimization using bio inspired localization algorithm: A survey. *Int. J. Curr. Eng. Sci. Res.* 10(4), 74–80.

38. Singh, P, Khosla, A Kumar, Khosla, A (2017) A novel approach for localization of moving target nodes in wireless sensor networks. *Int. J. Grid Distrib. Comput.* 10(10), 33–44.

39. Roumeliotis, S I, Bekey, G A (2000) Bayesian estimation and Kalman filtering: A unified framework for mobile robot localization. In: *IEEE International Conference on Robotics and Automation*, vol. 3, pp. 2985–2992.

40. Reina, G, Vargas, A, Nagatani, K Yoshida (2007) Adaptive Kalman filtering for GPS-based mobile robot localization. In: *IEEE International Workshop on Safety, Security and Rescue Robotics (SSRR)*, pp. 1–6.

41. Patwari, N, Ash, J N (2005) Locating the nodes cooperative localization in wireless sensor networks. *IEEE Sig. Process. Mag.* 22(4), 54–69.

42. Jiang, J, Han, G, Zhu, C Dong, Zhang, Y (2011) Secure localization in wireless sensor networks: A survey. *JCM* 6(6), 460–470.

43. Kumar, A, Khosla, A, Saini, J S Singh (2012) Meta-heuristic range based node localization algorithm for wireless sensor networks. In: *International Conference on Localization and GNSS (ICL-GNSS)*, pp. 1–7.

44. Kumar, A, Khosla, A, Saini, J S, Sidhu (2015) Range-free 3d node localization in anisotropic wireless sensor networks. *Appl. Soft Comput.* 34, 438–448.

45. Lee, S, Park, C, Lee, M J, Kim S (2014) Multihop range-free localization with approximate shortest path in anisotropic wireless sensor networks. *EURASIP J. Wirel. Commun. Netw.* 1, 1–12.

46. Awad, A, Frunzke, T, Dressler F (2007) Adaptive distance estimation and localization in WSN using RSSI measures. In: *10th Euromicro Conference on Digital System Design Architectures, Methods and Tools*, pp. 471–478.

47. Savvides, A, Park, H Srivastava, M B (2002) The bits and flops of the n-hop multilateration primitive for node localization problems. In: *Proceedings of the 1st ACM International Workshop on Wireless Sensor Networks and Applications*, pp. 112–121.

48. Savvides, A, Han, C C, Shrivastava (2001) Dynamic fine-grained localization in ad-hoc networks of sensors. In: *7th Annual International Conference on Mobile Computing and Networking*, pp. 166–179.

49. Sumathi, R, Srinivasan, R (2011) RSS-based location estimation in mobility assisted wireless sensor networks. In: *6th International Conference on Intelligent Data Acquisition and Advanced Computing Systems (IDAACS)*, vol. 2, pp. 848–852.

50. Guo, Z, Guo, Y Hong, Jin, F, Feng, Y, Liu, Y, Feng, Y, Liu, Y (2010) Perpendicular intersection: Locating wireless sensors with mobile beacon. *IEEE Trans. Veh. Technol.* 59(7), 3501–3509.

51. Kim, K, Lee, W M B A L (2007) a mobile beacon-assisted localization scheme for wireless sensor networks. In: *16th International Conference on Computer Communications and Networks*, pp. 57–62.

52. Karim, L Nasser, El Salti, N (2010) RELMA: A range free localization approach using mobile anchor node for wireless sensor networks. In: *IEEE IEEE Global Telecommunications Conference (GLOBECOM 2010)*, pp. 1–5.

53. Li, H Wang, Li, J X H (2008) Real-time path planning of mobile anchor node in localization for wireless sensor networks. In: *International Conference on Information and Automation*, pp. 384–389.

54. Chen, Y S, Ting, Y J, Ke, C H, Chilamkruti, N, Park, J H (2013) Efficient localization scheme with ring overlapping by utilizing mobile anchors in wireless sensor networks. *ACM Trans. Embedded Comput. Syst. (TECS)* 12(2), 20.

55. Khelifi, M, Benyahia, I, Moussaoui, S, Abdesselam, Naït (2015) An overview of localization algorithms in mobile wireless sensor networks. In: *International Conference on Protocol Engineering (ICPE) and International Conference on New Technologies of Distributed Systems (NTDS)*, pp. 1–6.

56. Stone, K, Camp, T (2012) A survey of distance-based wireless sensor network localization techniques. *Int. J. Pervasive Comput. Commun.* 8(2), 158–183.

57. Huang, S C, Chang, H Y (2015) A farmland multimedia data collection method using mobile sink for wireless sensor networks. *Multimedia Tool. Appl.* 2015, 1–16.

58. Gholami, M, Cai, N, Brennan, R (2013) An artificial neural network approach to the problem of wireless sensors network localization. *Robot. Comput. Integr. Manufact.* 29(1), 96–109.

59. Singh, P, Saini, H S (2015) Average localization accuracy in mobile wireless sensor networks. *J. Mob. Syst. Appl. Serv.* 1(2), 77–81.

60. Wang, J, Han, T (2011) A self-adapting dynamic localization algorithm for mobile nodes in wireless sensor networks. *Procedia Environ. Sci.* 11, 270–274.

61. Ding, Y Wang, Xiao, C L (2010) Using mobile beacons to locate sensors in obstructed environments. *J. Parallel Distrib. Comput.* 70(6), 644–656.

62. Chen, H, Gao, F, Martins, M Huang, Liang, P J (2013) Accurate and efficient node localization for mobile sensor networks. *Mob. Netw. Appl.* 18(1), 141–147.

63. Wang, Y, Wang, Z (2011) Accurate and computation-efficient localization for mobile sensor networks. In: *International Conference on Wireless Communications and Signal Processing (WCSP)*, pp. 1–5.
64. Shi, Q, Huo, H, Fang, T, Li, D (2009) A 3d node localization scheme for wireless sensor networks. *IEICE Electron. Express* 6(3), 167–172.
65. Wang, L Zhang, Cao, J D (2012) A new 3-dimensional DV-Hop localization algorithm. *J. Comput. Inf. Syst.* 8(6), 2463–2475.
66. Xu, Y Zhuang, Gu, Y J J An improved 3d localization algorithm for the wireless sensor network.
67. Li, J Zhong, Lu, X I T (2014) Three-dimensional node localization algorithm for WSN based on differential RSS irregular transmission model. *J. Commun.* 9(5), 391–397.
68. Ahmad, T, Li, X J, Seet, B C (2017) Parametric loop division for 3d localization in wireless sensor networks. *Sensors* 17(7), 1697.
69. Zhang, L Zhou, Cheng, X Q (2006) Landscape-3d; a robust localization scheme for sensor networks over complex 3d terrains. In: *Proceedings of the 31st IEEE Conference on Local Computer Networks*, pp. 239–246.
70. Tan, G Jiang, Zhang, H, Yin, S, Kermarrec, Z, AM (2013) Connectivity-based and anchor-free localization in large-scale 2d/3d sensor networks. *ACM Trans. Sens. Netw. (TOSN)* 10(1), 6.
71. Peng, L J, Li, W W (2014) CCDC) The improvement of 3d wireless sensor network nodes localization. In: *The 26th Chinese Control and Decision Conference*, pp. 4873–4878.
72. Zhang, B, Fan, J, Dai, G, Luan, T H A hybrid localization approach in 3d wireless sensor network. *Int. J. Distrib. Sens. Network* 2015, 1–11.
73. Yu, G, Yu, F, Feng, L (2008) A three dimensional localization algorithm using a mobile anchor node under wireless channel. In: *IEEE International Joint Conference on Neural Networks (IEEE World Congress on Computational Intelligence)*, pp. 477–483.
74. Liu, L Zhang, Shu, H, Chen, J (2011) A RSSI-weighted refinement method of IAPIT-3d. In: *International Conference on Computer Science and Network Technology (ICCSNT)*, vol. 3, pp. 1973–1977.
75. Achanta, H K, Dasgupta, S (2012) Ding Optimum sensor placement for localization in three dimensional under log normal shadowing. In: *5th International Congress on Image and Signal Processing (CISP)*, pp. 1898–1901. IEEE.
76. Teymorian, A Y, Cheng, W, Ma, L, Cheng, X, Lu, X 3d underwater sensor network localization.
77. Cheng, W, Teymorian, A Y, Cheng, Ma L, Lu, X, Lu, X Z (2008) Underwater localization in sparse 3d acoustic sensor networks. In: *INFOCOM 2008. The 27th Conference on Computer Communications*, pp. 236–240. IEEE.
78. Zhou, Z, Cui, J H, Zhou, S (2007) Localization for large-scale underwater sensor networks, Networking 2007. In: *Ad hoc and Sensor Networks, Wireless Networks, Next Generation Internet*, pp. 108–119.
79. Tomic, S, Beko, M, Dinis (2017) 3-d target localization in wireless sensor networks using RSS and AOA measurements. *IEEE Trans. Veh. Technol.* 66(4), 3197–3210.
80. Chan, Y T Chan, Read, F, Jackson, W, Lee, B R (2014) Hybrid localization of an emitter by combining angle-of-arrival and received signal strength measurements. In: *IEEE 27th Canadian Conference on Electrical and Computer Engineering (CCECE)*, pp. 1–5.
81. Xiang, Z, Ozguner, U (2005) A 3d positioning system for off-road autonomous vehicles. In: *IEEE Intelligent Vehicles Symposium Proceedings*, 130–135.
82. Yu, K 3-d localization error analysis in wireless networks. *The IEEE Transactions on Wireless Communications* 6(10), 3473–3481.

83. Tarrío, P, Besada, J A, Casar, J R (9–12 July 2013) Fusion of RSS and inertial measurements for calibration-free indoor pedestrian tracking. In: *Proceedings of the 16th International Conference on Information Fusion*, Istanbul, Turkey, pp. 1458–1464.

84. Wang, H, Lenz, H, Szabo, A, Bamberger, J, Hanebeck, U D (2007) *WLAN-Based Pedestrian Tracking Using Particle Filters and Low-Cost MEMS Sensors*, pp. 1–7.

85. Woodman, O, Harle, R (21–24 September 2008) Pedestrian localization for indoor environments. In: *Proceedings of the International Conference on Ubiquitous Computing (UbiComp '08)*, New York, NY, pp. 114–123.

86. Shu, J Zhang, Liu, R, Wu, L, Zhou, Z, Zhou, Z (2009) Cluster-based three-dimensional localization algorithm for large scale wireless sensor networks. *JCP* 4(7), 585–592.

87. Kennedy, J (2011) *Particle Swarm Optimization. Encyclopedia of Machine Learning*, pp. 760–766, Springer.

88. Zhang, X Wang, Fang, T J (2014) A node localization approach using particle swarm optimization in wireless sensor networks. In: *International Conference on Identification, Information and Knowledge in the Internet of Things (IIKI)*, pp. 84–87.

89. Yang, X S (2009) Firefly algorithms for multimodal optimization. In: *International Symposium on Stochastic Algorithms*, pp. 169–178, Springer.

90. Harikrishnan, R, Kumar, V J S, Ponmalar, P S (2016) Firefly algorithm approach for localization in wireless sensor networks. In: *3rd International Conference on Advanced Computing, Networking and Informatics*, pp. 209–214, Springer.

2 Literature Survey on Data Aggregation Techniques Using Mobile Agent and Trust-Aware in Wireless Sensor Network

Neelakshi Gupta, Tripti Sharma, and Nitin Mittal

CONTENTS

2.1   INTRODUCTION

In recent research into Wireless Sensor Networks (WSNs), trust-aware data aggregation using a mobile agent is the new challenge. This network is made up of small sensor nodes, which contribute to the communication of data [1]. WSN is application specific, therefore, it has the ability to deploy thousands of small sensor nodes, which can assemble themselves and are capable of collecting data from their surroundings and route the data to the sink node by either single hops or multiple hops [2].

Sensor nodes constitute different units, known as the communication unit, sensor unit, processor unit, and energy unit, as shown in Figure 2.1. Sensor units collect data from their surroundings and send them to the processor unit. In the processor unit, data are processed and saved for future use. The processed data from the processor unit are then forwarded to the communication unit, which is accountable for communicating the particulars from node to node. The main job of the energy unit is to keep a check on residual energy in the nodes. A sensor node has limited onboard storage, processing, power, and radio capacities, due to its limited size. Therefore, WSNs necessitate efficacious mechanisms to utilize limited resources and to resolve associated problems. Routing is one of these mechanisms that elongates the network lifespan by lessening energy consumption in communication. Figure 2.1 shows the architecture of the sensor node.

WSN is widely used in many fields, such as precision agriculture, environmental applications, smart transportation, health care applications, smart buildings, military applications, and many more. When a sensor network is used for military applications, continuous monitoring is essential to keep a check on movement of the enemy "like" mass destruction security assistance, etc. WSN is also very useful in health applications as it can remotely monitor the patients and checks their heart rate and blood pressure, transmitting this information to a WSN doctor. Table 2.1 summarises the applications of WSN.

Energy saving is a challenging task for researchers, as more energy is consumed during data transmission. The consumption of energy also depends on other factors, like the redundancy of transmitted data, traffic flow of data packets, and communication bandwidth [18]. In WSNs, sensor nodes are not able to send data directly to the sink node, as fewer nodes lie at some distance from the sink. If they do so, then nodes which lie far away from the sink will die out earlier, as compared with the other nodes. Therefore, multi-hop communication is used, where all nodes send their data to their next-nearest node till they reach the sink node. The process of collecting and transmitting data from the nodes to the sink node is called the data aggregation process. Data aggregation is an important process, because its aims are to collate data from different sources. Different approaches for data aggregation include *tree-based*, *multipath*, *cluster-based*, *hybrid*, *centralized*, and *in-network data aggregation* approaches, which are studied with respect to different performance parameters,

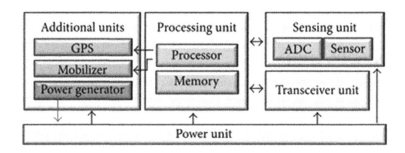

FIGURE 2.1 Architecture of sensor node.

TABLE 2.1

Summary of Various WSN Applications with Their Characteristics

Application	Characteristics
Precision agriculture [3]	Data under observation include humidity, temperature, soil depth, crop information, soil information, etc.
Environmental applications [4–7]	Information about coal mining, air pollution, forest fire detection, cyclones, earthquakes, volcano eruption, water quality parameters (pH levels, ammonia concentrations, and water levels)
Smart transportation [8–10]	Monitors air traffic, traffic congestion areas, targeting vehicles that are violating traffic rules.
Health care applications [11–12]	Monitors the fitness level and exercise level of the patient. It is used in different biomedical applications, such as continuous ECG monitoring.
Smart buildings [13–15]	Checking on different functionalities, like household appliances, biometric applications, and home security
Military applications [16, 17]	Monitors the movement of enemies at border areas, nuclear detection monitoring, secret communications in military applications.

such as energy efficiency, data accuracy, latency, overhead, network lifetime, and scalability of the sensor network.

Research has pursued data aggregation techniques to lower the energy consumption in WSNs. In the past few years, protocols like Low-Energy Adaptive Clustering Hierarchy (LEACH), Power-Efficient Gathering in Sensor Information System (PEGASIS) and many more have been proposed by researchers to boost energy efficiency and to extend the network lifetime. For increasing the scalability of networks efficiently, hierarchical architecture is the preferred one. In this protocol, the holistic network is divided into layers, with nodes of an individual layer playing the same role. Clustering-based protocols are preferred, because they are energy-efficient methods. In clustering, there are two main parts: cluster members and a cluster head (CH). The selection of the CH depends on the estimated energy left in the node. The main responsibility of a CH is to remove redundancy and make transmission successful by reducing data traffic. The CH aggregates data from its cluster members and transmits it to the base station (BS). Thus, the process of clustering is divided into three different steps: selection of CH, formation of cluster, and data transmission.

The main drawback of clustering is the random distribution of the CH and the greater energy consumption by the CH, using single-hop communication. In addition, unequal load balancing occurs during data transmission in multi-hop communication with the BS. An alternative approach to gathering information from the nodes of network is with the help of a mobile agent (MA). Data aggregations with the help of MAs satisfy the requirement of energy efficiency and network lifetime. MA is a self-adaptive type of middleware technology, which moves autonomously in a network without any pre-installed requirements.

The use of MA in the context of WSN has various applications, like data aggregation, target tracking, reprogramming, healthcare, image querying, and intrusion detection, etc. When the MA is dispatched from the BS, it follows one of two itineraries, namely single itinerary planning (SIP) or multi-itinerary planning (MIP). Different algorithms of SIP have been compared and their findings and disadvantages will be discussed in this chapter. Various MIP algorithms, which describe the number of MAs required on the basis of partition, will also be discussed.

Further research has been pursued on external and internal attacks. Two principal methods of providing security in WSN have been proposed, namely cryptography and intrusion detection system (IDS). Cryptography is well suited to foreign attacks, but it fails to manage domestic attacks. IDSs are used to reveal malicious nodes in the network. IDS is a complex process, which follows a certain set of norms, with which the results should be examined. Thus, to ensure the security of WSN, trust-aware management has been introduced over the past few years. It calculates the value of trust of the nodes, which is further used in different processes of WSN, such as data aggregation and data routing. A detailed study of trust-aware management is discussed in the following sections.

The remaining sections of this chapter are structured as follows. A detailed survey of routing protocols is described under Section 2.2. Various data aggregation and clustering techniques are described in Sections 2.3 and 2.4. The drawbacks of clustering and the introduction of MA, along with their itineraries, are given in Sections 2.5 and 2.6. The importance of trust in WSN is discussed in Section 2.7. In the Conclusions, different research strategies are compared and discussed.

2.2 ROUTING PROTOCOLS IN WSN

Different routing protocols are described in this section. Routing is the foremost parameter in the design of WSN. Researchers classify routing protocols into four different categories: network structure, communication model, topology-based protocols, and reliable routing protocols. Figure 2.2 shows the detailed taxonomy of routing protocols [19].

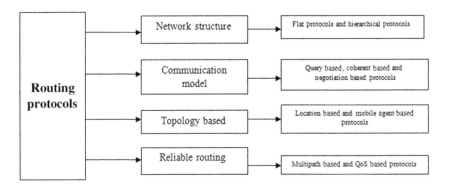

FIGURE 2.2 Taxonomy of routing protocols [19].

2.2.1 NETWORK STRUCTURE

The network structure can be defined in accordance with the distribution of the sensor nodes. It can also be defined on the basis of the type of connection between the nodes, and by the routing method used to transfer the information. It can be classified as follows:

a) *Flat Routing Protocols.* In flat-based protocols, all nodes of the sensor network are at the same level within it. It is a multi-hop routing technique [20] (Table 2.2).

b) *Hierarchical Routing Protocols.* The whole network is divided into a small group of nodes called clusters, in which the CH is chosen, depending on the residual energy. When clusters are created in a sensor network, it results in an extended network lifetime, and improved energy efficiency and stability [31]. Different hierarchical routing protocols have been studied, and they are compared with respect to different parameters in Table 2.3.

2.2.2 COMMUNICATION MODEL

In these protocols, more data can be transferred within the energy available in the node. The protocols for this category can also achieve the theoretical optimum in point-to-point and broadcast networks with respect to dissemination rate and energy use. The controversy associated with this communication model is that the data transmission ratio is not high. Therefore, the delivery of data to the destination is uncertain.

i) *Query-Based Protocols.* In this category, destination or sink node will broadcast an inquiry to find specific events among WSN. A particular node, which meets the query, will respond to the destination node from where the query was generated.

TABLE 2.2

Comparison of Various Flat Routing Protocols

Routing Protocol	Classification	Data Aggregation
SPIN [21]	Negotiation	Yes
DD [22]	Data-centric	Yes
Rumour routing [23]	Data-centric	No
Gradient-based routing [24]	Data-centric	Yes
Constrained Anisotropic Diffusion Routing (CADR) [25]	Data-centric	No
Energy-Aware Routing (EAR) [26]	Data-centric	Yes
Minimum Cost-Forwarding Algorithm (MCFA) [27]	Data-centric	No
Active Query-Forwarding In-Sensor network (ACQUIRE) [28]	Data-centric	Yes
Sequential Assignment Routing (SAR) [29]	QoS-based	Yes
Energy-Aware Data-Centric Routing (EAD) [30]	Data-centric	Yes

TABLE 2.3

Comparison between Hierarchical Routing Protocols

Routing Protocol	Energy Efficiency	Cluster Stability	Classification	Network Lifetime
LEACH [32]	Low	Moderate	Distributed	High
TEEN [33]	High	High	Distributed	High
APTEEN [34]	Moderate	Very low	Hybrid	Moderate
PEGASIS [35]	Low	Very low	Chain-based	High
HEED [36]	Moderate	High	Distributed	High
CSPEA [37]	High	Moderate	Distributed	Moderate
SEP [38]	Moderate	Moderate	Distributed	Moderate
DWEHC [39]	Very high	High	Distributed	High
EECHE [37]	High	Moderate	Distributed	Moderate
CCS [40]	Low	Low	Distributed	High
DEEC [41]	High	High	Distributed	Moderate
HGMR [42]	Moderate	High	Location-based	Moderate
GAF [43]	Moderate	Moderate	Location-based	High
BCDCP [44]	Low	High	Centralised	Moderate

ii) *Coherent-Based Protocols.* Data are transported after applying the data processing techniques, such as removing redundancy. Coherent processing is performed on the data to achieve energy-efficient routing.

iii) *Non-Coherent Protocols.* In these protocols, the nodes process raw data themselves before forwarding them for further processing within the network.

iv) *Negotiation-Based Protocols.* These protocols use metadata negotiations for reducing redundant network transmissions.

2.2.3 Topology-Based Protocols

These are based on topology with all nodes of the network in three ways, e.g., reactive, proactive and hybrid protocol. The protocols used in this category are described as follows:

i) *Location-based Routing Protocols.* When using information about the position of nodes within the network, nodes are distinguished. The CH aggregates data from all its cluster members, removes the redundancy, and transmits the data into the BS.

ii) *Mobile Agent-Based Routing Protocols.* MA is a middleware technology which migrates within the network for the collection of data. It works autonomously and intelligently. It overcomes the traditional client–server model.

2.2.4 RELIABLE ROUTING PROTOCOLS

These protocols are more resistant to network failures as a result of either realisation of load balancing routes or of satisfactory specific Quality of Service (QoS) metrics, such as delay, energy, and bandwidth. The protocols defined for this category are as below:

i) *Multipath-Based Protocols.* They prefer to use multiple paths in comparison with using a single path, because, when a particular route has failed, it can opt for an alternative path. These protocols will increase the robustness and fault tolerance of the network.

ii) *QoS-Based Protocols.* The network must balance energy utilisation with data quality. Each time a sink node requires data from nodes of a sensor network, the transmission must meet explicit quality levels. Also, it should take care of QoS parameters, such as delay, bandwidth, energy, etc.

2.3 DATA AGGREGATION APPROACHES

Energy utilisation is the prime concern in sensor networks, as sensor nodes cannot send data to the BS directly. Sometimes, redundancy is increased by the neighbouring sensor nodes, which results in congestion at the BS, which becomes loaded with enormous amounts of data. As a consequence, it is necessary that redundancy is removed at some intermediary levels, which can result in a reduction in the number of data packets, which are forwarded to the BS. Reducing traffic at the BS saves energy and bandwidth. This approach is known as data aggregation [45]. It increases the lifetime of the network by decreasing the total data transmissions. The multi-hop approach is used to forward the sensed data to the CH. The CH uses either a traditional wired network or a fixed wireless medium to deliver the aggregated data to the BS. There are various approaches to data aggregation: the tree-based approach, centralised approach, in-network approach, cluster-based approach, and the multi-path approach.. Figure 2.3 describes the various data aggregation approaches.

Clustering is the most efficient data aggregation approach. It extends the network lifetime by reducing energy consumption. With the help of clustering, direct interaction among sensor nodes and the BS is decreased, which results in removed redundancy and reduced energy consumption.

2.4 CLUSTERING IN WSN

WSN is application oriented but, still, these networks suffer from several restrictions like energy constraints, limited battery capacity, etc. Nodes in the sensor network have less power and range, so cannot communicate directly with a sink node, if it is located at some distance away. If a sensor node tends to communicate directly with a distant sink node, it will give rise to greater use of energy resources, which results in network failure. Thus, to boost the lifetime of the network, the clustering technique is used. The selection of CH depends on the energy resources and it coordinates

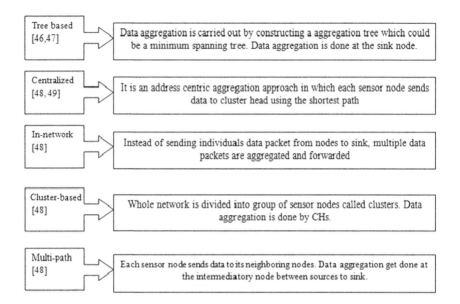

FIGURE 2.3   Data aggregation approaches.

activities inside and outside the cluster. It also removes redundancy from the data gathered from other cluster members.

A number of clustering schemes have been proposed in recent years. These schemes inform the selection of CHs and the aggregation of data into clusters with network design. To reduce energy consumption of the sensor node during communication, various methods have been proposed. Clustering is defined as one of the main approaches to the development of protocols, which are effective, efficient, and scalable in WSNs. In clustering, the sensors of a network are grouped into small clusters, and each cluster is earmarked to a CH. After a certain interval of time, sensor nodes send the data to the CH, and it is the role of the CH to forward the data to the BS. Some benefits of clustering are listed below:

1. Clustering sustains the bandwidth for communication.
2. Clustering can balance the level of topology of the sensor network and reduce overhead traffic, with decreased overall maintenance costs.
3. It controls repetition of the substituted information between the nodes.
4. It helps to extend the life of the network.
5. Energy utilisation is decreased by management of the sleep cycle.
6. Fewer data packets are sent over the network.
7. Clustering prevents resending, reduces data repetition in the covered domain, and collisions are avoided.

A popular routing-based protocol based on clustering is LEACH, which is subdivided into rounds; in each respective round, the CH role is circulated between nodes, to maintain balance. With the intention of optimising energy use in the network, nodes are selected in terms of CH circularity and randomness. Based on the principle of proximity, normal nodes, which are referred to as cluster members, are a part of the corresponding CH. The main drawback of LEACH is that the CH selection is based upon probability, rather than the surplus energy of the nodes. Therefore, sometimes the wrong CH is selected, which has a small amount of energy, and, as a result, the CH dies out early. Table 2.4 shows a comparison of different parameters of LEACH and its variants [46–49]. The LEACH variants exhibit improved functioning than the standard LEACH protocol.

2.5   MOBILE AGENT-BASED CLUSTERING

Earlier research proposed clustering as a prime technique for achieving energy efficiency in WSN. This is because clustering allows the sensor nodes to gather the data at a single node, the CH, which forwards the data to BS, rather than allowing every single sensor node to send the data to the BS. The clustering evolves from the LEACH protocol, which uses three phases: CH selection, the formation of the cluster, and data forwarding *via* the CH. Clustering, in itself, suffers drawbacks, like random distribution of CH, greater energy consumption by the CH in single-hop communication with the BS, or unequal load balancing in multi-hop communication with the BS. Another approach to aggregate information from the sensor nodes is with the help of a mobile agent (MA) [59].

MA can migrate within technology that can migrate within a sensor network, and perform the assigned task such as data aggregation, data fusion, etc. It replaces the traditional client–server approach because of its different benefits. Table 2.5 shows a comparison between the traditional–client server model and the mobile agent model.

The MA has great advantages, such as short time of execution. This is because each and every node contributes effort to data aggregation. When the MA dispatched by the server in the network, it visits and collects interesting data, and aggregates it. The aggregated data is forwarded to the server by the MA. A number of MA-based networks have been developed by various researchers.

2.6   MA ITINERARIES IN WSNS

There are different uses of MA in the context of WSN, for different applications, such as data aggregation, reprogramming, healthcare, intrusion detection, etc. The performance of all these applications is entirely dependent on the route followed by an MA. The itineraries for MA can be designed by either choosing single itinerary or multiple itineraries.

2.6.1   SINGLE ITINERARY PLANNING

A single agent is dispatched from the BS, when using Single Itinerary Planning (SIP). Two different heuristic algorithms have been proposed [60]: (i) Local Closest First (LCF), which looks at the subsequent node with the shortest distance from the

TABLE 2.4
Comparison of LEACH and its Variants

Protocol	Networking Approach	Mobility	Network Lifetime	Scalability	Energy Consumption
LEACH [32]	Proactive/clustering	No	Higher	Good	Maximum
LEACH-C [50,51]	Proactive/clustering	No	Higher	Good	Minimum
LEACH-F [52]	Proactive/clustering	No	Higher	Limited	Minimum
LEACH-B [53]	Proactive/clustering	No	Higher	Good	Minimum
TL-LEACH [54]	Proactive/clustering	No	Higher	Very good	Minimum
LEACH-E [55]	Proactive/clustering	No	Higher	Very good	Minimum
MH-LEACH [54]	Proactive/clustering	No	Higher	Very good	Minimum
LEACH-M [56]	Proactive/clustering	Yes	Average	Good	Minimum
I-LEACH [55]	Proactive/clustering	No	Higher	Very good	Minimum
LEACH-A [52]	Proactive/clustering	No	Average	Good	Maximum
Cell- LEACH [57]	Proactive/clustering	No	Average	Very good	Minimum
V-LEACH [58]	Proactive/clustering	No	Higher	Very good	Maximum

TABLE 2.5

Comparison between Traditional Client–Server Model and Mobile Agent Model

Traditional Client–Server	Mobile Agent-Based
Congestion of traffic	Reduces network load
Hot-spot problem	Mobility of server/sink
Not flexible according to application	Application oriented
More energy usage	Less consumption of energy
Lifetime of the network is shortened	Enhanced network lifetime
Not dynamic	Dynamic
Not scalable	Scalability is possible
Less reliable	More reliable

current node; and (ii) Global Closest First (GCF), which looks at the node which is next to the sink. A third routing protocol is the MA-based directed diffusion protocol (MADD) [61]. MADD is near-identical to LCF, and they are differentiated only on the basis that, in MADD, the MA selects the node which is farthest from the sink as the first node. Hence, these three routing protocols (LCF, GCF, and MADD) are not complicated to follow, but are not capable of being upscaled, because the distance between the source node and destination node determines the route.

In [62], two algorithms were designed to accomplish energy-efficient itineraries, namely the itinerary designed for minimum energy for first-source selection (IEMF), and the itinerary designed for minimum energy algorithm (IEMA). IEMF follows the round-robin method for tentatively by choosing the first source node to be visited according to estimated communication cost. Then, the LCF algorithm is applied to the rest of the source nodes. Such a process proposes various itineraries for candidates, each route contributing to energy costs. Afterwards, the IEMF selects the route with the lowest energy cost. IEMA is an iterative version of IEMF, i.e., IEMA defines the sequence by which the unused source nodes are visited as the first source node. Rather than searching for global network information, LCF looks for the adjacent MA hop, which depends on the MA's current position. In addition, LCF, GCF, IEMF, and IEMA algorithms are developed using a SIP with degraded conduct in a wide network.

2.6.2 MULTI-ITINERARY PLANNING

SIP performs well for a small-scale network but it fails in large-scale networks, because of the following drawbacks:

i) As there is only one MA, which has to take a tour of hundreds of nodes within the network for data collection, this network suffers from long delays.

ii) There is a possibility that MA can be lost during migration to several nodes.

iii) The size of the data packet is increased when it aggregates from node to node and consumes more energy.

iv) As MA accumulates more and more data, its reliability decreases.

MIP overcomes these drawbacks by sending multiple MAs. Thus, MIP can be defined as an itinerary in which more than one MA is dispatched by the server in the network. The determination of the best possible route for MA is inspired by the arrangement of a global network in a large-scale network in which the MIP approach takes place. For calculations of the required number of Mas, along with their paths, the near-optimal itinerary design (NOID) algorithm was introduced in [63]. The distance between source nodes is the main parameter used in cost–weight calculations. This distance is calculated based on the minimum spanning tree (MST). Thus, it acts as a compromise to balance, so that the MA first visits a source node with inadequate energy (where the MA data are less).

However, the BST–MIP algorithm is differentiated from the NOID due to the fact that, at some stage in weight calculation in a totally connected graph (TCG), which indicates that all source nodes in a particular sub-tree should be considered as a group, it makes use of a balance factor [64]. The balance factor is used to get a flexible control between energy costs and task duration. A genetic algorithm (GA), based on MIP, was designed to be used to calculate the desired number of MAs, along with their MIP itineraries [65].

The idea behind GA-MIP is to determine the required number of MAs and their allocated source nodes, with the help of a two-tier coding method, based on GA. This coding acts as a gene that involves a series array (sequence) and a set array (group). The series array accommodates segments corresponding to a number of source nodes, associated with a MA. Each number of source arrays corresponds to source nodes for each segment. Each iteration uses two basic GA operations (crossover and mutation) and a fitness function for the selection of better genes. Despite good delay and energy consumption performances, GA-MIP also has a high computational complexity because of the need to maintain global network information in each iteration. In this GA-MIP referred to above, the geographical information of nodes and the cost of the MA itinerary are the main parameters which are used to determine the itineraries for the optimal number of MAs.

The greatest information in the greater memory based MIP (GIGM-MIP) algorithm takes geographical information into account in order to define useful number of MAs, along with their itineraries and data size in each partition [66]. The k-means clustering in GIGM-MIP tends to split the network into different clusters (partitions). The GIGM-MIP algorithm considers the data size of the source nodes for each cluster after partitioning the network. This data size determines the number of MAs allocated to a particular partition, so that each cluster is allocated more than one MA. The node with the largest free payload data is then assigned to the MA which has the maximum data size in comparison with the other nodes. This process is reiterated until all the MAs have almost the same size of payload data.

With this solution, a balance is maintained between the transported data and the distributed MAs, and it leads to a decrease in energy consumption. To find out the desired number of MAs and their itineraries, an immune-inspired algorithm is proposed, which is also called the Clonal Selection Algorithm for MIP (CSA-MIP) [67]. It uses the same two-level method of coding as the GA-MIP method. The difference between CSA-MIP and the previous MIP is that it uses an evolutionary two-stage search process for identifying overall search capabilities, with a different assigned mutation operator at each stage. While using these search methods, the solutions obtained from it result in variability in the number of MAs. In addition, the imbalance is decreased when the desired number of sensor nodes is assigned to each MA. This situation may give rise to increased chances of achieving better solutions (Table 2.6).

2.7   TRUST-AWARE WSN

The MA-based data aggregation provides efficient and effective data aggregation, based on the MA itinerary. In addition to the migration of these energy-efficient MAs, another important matter of interest is the safety of MA in terms of malicious or compromised nodes. The topic of trust is an open and current field in WSN. A number of trust schemes have been surveyed and studied from different perspectives. When MA encounters a malicious node, a false reprogramming can be done by it which, in turn, results in corruption of the network, stopping the functioning of the MA, etc.

In the past few years, trust management has become included in the literature for assessing the node's trust value, based on the functional behaviour of the protocol. This assessed or calculated trust value can also be used for decision making with respect to data routing, data aggregation, and intrusion detection [68]. Two approaches to data aggregation in a trust-aware network are suggested in [69], where the first is the aggregation input collection and the second one is an inconsistency check. Data are aggregated at regular intervals of time, where the data are checked for inconsistency, if there is any need to calculate the trust of nodes.

Another secure algorithm for selecting the CH is achieved by calculating the weight of each node, with the intention of confirming that the minimum energy utilised is used for safe selection. The node's weight combines different metrics, including trust metrics (node behaviour), which assists the decision to select the CH, which means that the node will never be a malicious one [70].

The trust metric is conclusive and permits the proposed algorithm to prevent the selection of a CH by any malicious node in the area, even if the remaining parameters support it. Other node metrics are also studied, which includes waiting time, degree of connectivity, and the distance between the nodes. The CH selection, using the weights of members' nodes, is completed. MAEF is proposed, which is based on energy- and fault-conscious data aggregation in WSN. MA is sent to the selected CH, and this CH is chosen on the basis of the highest node density [71]. MA only visits selected CHs to collect data, which show improvements in network life, and also in energy consumption.

TABLE 2.6

Comparison of Different Algorithms of MA Itineraries

Algorithm	SIP/MIP	Itinerary Planning Method	Findings	Drawbacks
LCF	SIP	It indicates the next neighbourhood node with a smaller distance from the current node	Low complexity	Performance decreases with an increase in network size
GCF	SIP	The next node will be the node which is nearest to the sink	Low complexity	Performance degrades as the network size increases
MADD	SIP	The next node will be the farthest node from the sink	Low complexity.	Scalability is decreased.
IEMF	SIP	It adopts the round-robin method.	Improves the solutions derived by LCF	Higher computational complexity and low scalability.
IEMA	SIP	It indicates the visiting order for the rest of the nodes, including the first source node.	Improves the solutions by IEMF	Higher computational complexity and low scalability.
TSP	SIP	Examines k opt neighbourhoods of a given tour to derive a shorter one	A successful method for solving TSP	Non-trivial implementation
BST	MIP	Builds a minimum spanning tree in which one branch is assigned with one MA	Low computational complexity	Initially created branches absorb all the network.
NOID	MIP	Number of nodes within radius of processing element.	Reliable migration *via* energy-aware links.	Higher computational complexity
SNOID	MIP	Concentric zones are built around the sink to construct itineraries for MA.	Minimum cost while joining trees	Complexity increases.
TBID	MIP	Low-cost itineraries are designed to minimise total aggregation cost.	Reducing MA random walk	Costly in terms of updating tree network.
CL-MIP	MIP	Area to be visited by MA is decided by VCL	Efficiency is increased	Optimal value is not measured
AG-MIP	MIP	It uses the angle gap to divide the network	It reduces contention and interferences among MAs.	Optimal angle is difficult to determine
GA-MIP	MIP	Optimising links	As per requirement for best-suited route	Delay is increased
GIGM MIP	MIP	It is based on the balance between geographic information and data size.	Low complexity	Delay is increased.

2.8 SUMMARY AND CONCLUSIONS

Today, the WSNs have immensely expanded their vital role in data-efficient selection and transport of data to destinations. The utilisation of energy resources is a critical issue for the networks, particularly in wireless networks, which are characterised by shorter battery lives. The complexity and dependence of corporate operations performed on WSNs require efficient routing techniques and protocols to ensure network interlinking and information routing, with the least energy expenditure. In this paper, the attention is given to the energy-efficient protocols, associated with WSNs.

With the requirements of the next generation of wireless networks and personal communication systems, a new type of system, based on the MA, has been developed. MA offers a unique and promising solution to increase the efficiency of energy use while aggregating data, with this agent-based approach being the most promising one for solving the data problem of WSN.

This approach can easily eliminate redundancy in data, with only the relevant specific data being transferred to the sink. MA will visit the network for data aggregation by two types of itineraries planned by the BS. It can follow either the single itinerary planning or multi-itinerary planning route. This paper also discussed trust metrics, issues related to the construction of a WSN, and issues dealing with trust management. There is no standard model available which can be used in current trust systems to offer safety or resilience to attacks. The designers are trying to solve the problem of trustworthiness from various angles. Future trust management research will focus on a generalised, scalable, and reconfigurable model of trust, which is accessible to different distributed computing systems, which deal with malicious and non-malicious nodes, while processing the data. Such a development will improve security problems in a better way and can meet the application demands of the consumer. Table 2.7 summarises the literature related to data aggregation, clustering, mobile agents, and trust-aware.

2.9 FUTURE SCOPE

The relevant literature suggests some unique ideas, covering most of the areas of WSN, such as clustering, data aggregation, mobile agents, MA itineraries and trust-aware protocols. Many of the topics which are presented in this chapter are open for new discussions and for further research. This literature survey is intended for all who work with WSNs and related fields, to keep up with progress in the field. One of the most-important current topics in WSN is trust-aware, which needs to be explored in more detail. In trust-aware, a malicious node has to be detected by calculating the trust value before the data aggregation process is run. It is preferred over traditional cryptography methods because it has minimal computational cost.

TABLE 2.7
Survey Related to Data Aggregation, Clustering, Mobile Agent and Trust-Aware

Researcher	Clustering/ Data Aggregation/ Mobile Agent	Year	Findings
Bista et al. [72]	Data aggregation	2009	They proposed a balanced approach of data aggregation, which is energy efficient for WSNs, called the DP scheme. The DP scheme established and executed a round-robin set of paths so that all nodes can contribute to the collection and transmission of data to the sink. Energy dissipation, however, was enhanced.
Faheem and Boudjit [73]	Data aggregation	2010	They presented distributed sink location updates and the SN-MPR mobile sink data collection mechanism, which is a tree-based data aggregation approach. This mechanism used multi-point relays (MPRs) to update and query the sink location. However, data delivery was delayed.
Zhao and Yang [74]	Data aggregation	2010	They introduced joint SDMA technology design with multiple mobile collectors
Nazir and Hasbullah [75]	Data aggregation	2010	They addressed the problem of the hotspot and developed a mobile sink-based routing protocol (MSRP) as a solution, which extends the lifetime in a clustered WSN.
Yang et al. [76]	Data aggregation	2011	They have developed an approach to suppress aggregate data in-network, using order compression techniques.
S. Guo and Y. Yang [77]	Data aggregation	2012	They proposed a Data Collection Cost Minimisation framework (DaGCM) for simultaneous data upload, which is limited at all anchor points by flow conservation, connection capacity, sensor compatibility, and overall travel time of the mobile agent.
Xu et al. [78]	Data aggregation	2013	They present data aggregation framework to integrate a hierarchical order with multiple resolutions with CH for optimisation of the transmitted data.
Soltani et al. [79]	Data aggregation	2014	The proposed source efficiency data fusion approach for large-scale WSNs. Data fusion is used to detect a dropped number of nodes in the active network and to reduce network resource consumption.
Shiliang et al. [80]	Data aggregation	2015	They exploited a tradeoff between data quality and energy consumption, which was used to enhance the accuracy of data aggregation in the event of heterogeneous energy restrictions.

(Continued)

TABLE 2.7 (CONTINUED)

Survey Related to Data Aggregation, Clustering, Mobile Agent and Trust-Aware

Researcher	Clustering/ Data Aggregation/ Mobile Agent	Year	Findings
Awang and Agarwal [81]	Data aggregation	2015	Two different data aggregation mechanisms are proposed where collector nodes are specified opportunistically, and this does not depend on the global knowledge of geographical location.
Zhong et. al. [82]	Data aggregation	2017	They proposed a twice-partitioned algorithm which depends on centre points and divides the network into several points. Moreover, they also exploited a tradeoff between neighbour amount of energy and residual energy, which is proposed to gather zonal data.
Wan et al. [83]	Data aggregation	2019	They proposed a similarity-aware data aggregation technique with the help of fuzzy-c means in clustered WSN.
Chen et al. [84–87]	Mobile agent	2005, 2006, 2007, and 2011	They proposed an architecture of MA-based WSN in which the MA can carry data, process it, and transport it to the BS. MA is application specific and can adjust its behaviour depending on QoS requirements, e.g., data delivery, latency.
Chong et al. [88]	Mobile agent	2003	This team identifies that its application is limited to military use. They also developed MEMS in which they enhance the MA applications in an embedded system.
C.-L. Fok et al. [89]	Mobile agent	2005	This team presented Agilla which is a type of MA middleware which can rapidly increase its application deployment in WSN.
Sethi et al. [90]	Mobile agent	2011	They solved the problem of balancing the energy during data collection. They proposed energy-balanced cluster-routing, based on MA, in which cluster structure depends on cellular topology where the energy is balanced in inter-clusters and intra- clusters.
Aiello et al. [91]	Mobile agent	2011	They proved that MA can efficiently work in WSN- based structural health monitoring applications on large aircraft structures.

(Continued)

TABLE 2.7 (CONTINUED)

Survey Related to Data Aggregation, Clustering, Mobile Agent and Trust-Aware

Researcher	Clustering/ Data Aggregation/ Mobile Agent	Year	Findings
Chatterjee et al. [92]	Clustering	2002	They suggested WCA (Weighted Clustering Algorithm), which takes different mobile node parameters such as transmission power, mobility, and mobile node battery power. The main purpose of WCA is that it can handle CH ideally.
Xiu et al. [93]	Clustering	2016	They proposed a protocol for routing in WSN which depends on the monitoring of energy-balanced partition clustering (EBPC), in which data fusion rate nodes are controlled and intra-cluster communication, with a minimum path selection coefficient, is carried out.
Singh et al. [94]	Clustering	2016	They compared the energy efficiency of three protocols, viz. LEACH, FAIR, and SEP, in a heterogeneous environment.
Wu et al. [95]	Clustering	2016	They proposed a model which can control CHs efficiently in real-time, using heterogeneous sensors. They also proposed a generic software which enhances the energy efficiency.
Yadav et al. [96]	Clustering	2018	They designed a clustering algorithm on distributed energy efficiency, in which CH decides which has more residual energy and average energy
Gupta et al. [97]	Mobile agent	2003	They proposed a heuristic data aggregation algorithm on cluster basis, which solves the network lifetime problem of large-scale networks
Biswas et al. [98]	Mobile agent	2008	They claimed that the BS dispatched a MA which migrates to each node and travels either a pre-planned itinerary or is planned on the fly and collects data from all sensor nodes. MA-based data gathering is more convenient than the client–server model because of the huge number of sensor nodes distributed
Zhong et al. [99]	Mobile agent	2018	They addressed the problem of the hotspot and proposed a mobile sink-based routing protocol (MSRP) for the prolongation of network life in clustered WSNs.
Xing et al. [100]	Mobile agent	2008	They proposed rendezvous planning with mobile elements in WSN. In this planning, a rendezvous-based approach was discussed for the use of MEs to collect sensor data within time limits

(Continued)

TABLE 2.7 (CONTINUED)

Survey Related to Data Aggregation, Clustering, Mobile Agent and Trust-Aware

Researcher	Clustering/ Data Aggregation/ Mobile Agent	Year	Findings
Zhao et al. [101]	Mobile agent	2015	They provide the solution to improve the life of a large network in which the mobile sink collects data at regular intervals through predefined paths and each node of the network loads it towards the mobile sink through multi-hop communication.
Krishnan et al. [102]	Mobile agent	2015	They proposed a clustering architecture which is used to collect data efficiently in a TDMA time slot. The proposed algorithm increases the network life with low energy requirement.
Jose et al. [103]	Mobile agent	2015	They proposed an ABC and PSO bio-inspired technique, which satisfies the purpose of average packet delay and longevity.
Konstantopoulos et al. [104]	Mobile agent	2010	They calculated the residual energy of nodes of the clusters. In this technique, they described how MA is useful for balancing the energy between sensor nodes.
Deng et al. [105]	Trust-aware	2009	They suggested two different approaches of trust-aware, based on data aggregation, where data are aggregated at fixed intervals of time and then data are checked for inconsistency, if there is any need to calculate trust of nodes.
Rehman et al. [106]	Trust-aware	2017	They proposed a secure algorithm for selecting CH by calculating the weight of each node so as to confirm that the minimum energy utilised by the network is used for the secure selection. The node's weight is to combine different metrics, including trust metrics (sensor node behaviours), which assists a secure decision to select the CH, which means that the node will never be a malicious one
Wang et al. [107]	Trust-aware	2018	They proposed an Energy-Efficient Trust Management and Routing Mechanism for Software-Defined WSN. In this, the trust is evaluated at node level rather than at the controller level. During the data aggregation process, it also ensures the transmission of traffic.
Pati et al. [108]	Clustering	2017	They proposed a CH selection mechanism called an Energy-Efficient Approach, which works on ECHSA-1 and ECHSA-2 algorithms, where ECHSA-1 works with the Nash Equilibrium and ECHSA-2 works with the Sub-game Perfect Nash Equilibrium.

(Continued)

TABLE 2.7 (CONTINUED)

Survey Related to Data Aggregation, Clustering, Mobile Agent and Trust-Aware

Researcher	Clustering/ Data Aggregation/ Mobile Agent	Year	Findings
Hajji et al. [109]	Data aggregation	2018	They proposed multi-criterion-based Centralities Measures Routing Protocol (MCRP), which regulated congestion by favouring relay nodes. It is done by permanently choosing the relay nodes that have stable energy consumption and a balanced data emission and reception.
Khan et al. [110]	Clustering	2017	They proposed a Fuzzy-TOPSIS technique, the main purpose of which is to choose CHs efficiently on the basis of multicriterion decision making. Different parameters, like number of neighbour nodes, residual energy, and node energy consumption rate, are considered.
Gupta [111]	Data aggregation	2018	The author proposed an Improved Cuckoo Search-based meta-heuristic algorithm for energy-efficient clustering. This algorithm distributes the communication load uniformly to the CHs. Also, for the selection of the most energy-efficient route for data transportation from the CH to the sink, multi-hop routing is used.
Lingaraj [112]	Mobile agent	2017	Authors designed this Eagilla Mobile Agent middleware application to overcome the problem of waiting line delay. This new MA application organizes the WSN into multiple clusters to make the best use of available resources. It also supports the scalability and heterogeneity of large-scale networks.
Sasirekha and Swamynathan [113]	Mobile agent	2017	The authors used the clustering process of LEACH for dividing the cluster into groups. Data are aggregated by CH from its cluster members. MAs are used to aggregate data from the CH. This is an effective scheme in which the MA has to visit only CHs to collect the data, rather than collecting data from each sensor node. With this, it shortens the tour length for the MAs.

(Continued)

TABLE 2.7 (CONTINUED)
Survey Related to Data Aggregation, Clustering, Mobile Agent and Trust-Aware

Researcher	Clustering/ Data Aggregation/ Mobile Agent	Year	Findings
Cheng and Yu [114]	Data aggregation	2018	The authors proposed the Bounded Relay Combine-TSP-Reduce algorithm which is used to reduce the delay time of data aggregation. It used a path adjustment mechanism, which results in the shorter travelling path for the aggregation process.
Elshrkawey et al. [115]	Clustering	2018	Two different approaches are suggested by the authors to enhance the routing process of LEACH. One is a selection of proper CHs, and the other is to target those sensor nodes which are creating data congestion in the network. The proposed solution for CH selection is to modify the CH selection threshold value, whereas the solution for the second method solution is to use the TDMA schedule for data sending.
Ramluckun and Bassoo [116]	Clustering	2018	The authors combined PEGASIS and ACO to reduce delay and find the optimal path from where the transmission distance is minimal. Intra-cluster and inter-cluster communications are used to verify that minimum transmission energy is used.

BIBLIOGRAPHY

1. J.A. Stankovic (2008). Wireless sensor networks, *IEEE Computer*, 41(10), 92–95.
2. K. Yang (2014). Wireless sensor networks. In: *Principles, Design and Applications. Signals and Communication Technology*, Springer.
3. Xin Dong, Mehmet C. Vuran, and Suat Irmak (2013). Autonomous precision agriculture through integration of wireless underground sensor network with center pivot irrigation systems. In: *Ad Hoc Networks*, Elsevier, 1975–1987.
4. Vendula Hejlova and Vit Vozenilek (2013). Wireless sensor network components for air pollution in urban environment: Criteria and analysis for their selection, *Wireless Sensor Networks*, 5, 229–240.
5. Kavi K. Khedo, Rajiv Perseedoss, and Avinash Mungur (2010). A wireless sensor network Air pollution monitoring system, *International Journal of Wireless and Mobile Networks*, 2(2), 31–45.
6. B. Kechar, N. Houache, and L. Sekhri (2013). Using wireless sensor network for reliable forest fire detection, 3rd International conference on sustainable energy information technology, *Procedia Computer Science*,19, 794–801.
7. Yunus E. Aslan, Ibrahim Korpeoglu, and Ozgur Ulusoy (2012). A Framework for use of wireless sensor network in forest fire detection and monitoring, *Computers, Environment and Urban Systems*, 36(6), 614–625.
8. A. Pascale, M. Nicoli, F. Deflorio, B. Dalla Chaira, and U. Spagnolini (2012). Wireless sensor network for traffic management and road safety, *IET Intelligent Transportation Systems*, 6(1), 67–77.
9. Rahat Ali Khan, Shakeel Ahmed Shah, Muhammad Abdul Aleem, and Zulfiqar Ali Bhutto (2012). Wireless sensor networks: A solution for smart transportation, *Journal of Emerging Trends in Computing and Information Science*, 3(4), 566–571.
10. J. Guevara, F. Barrero, E. Vargas, and J. Becerra (2011). Environmental wireless sensor network for road traffic application, *IET Intelligent Transportation Systems*, 6(2), 177 –186.
11. Reza S. Dilmaghani, Hossein Bobarshad, M. Ghavami, S. Choobkar, and Charles Wolfe (2011). Wireless sensor networks for monitoring physiological signals of multiple patients, *IEEE Transactions on Biomedical Circuits and Systems*, 5(4), 347–356.
12. Joao Martinho, Luis Prates, and Joao Costa (2014). Design and Implementation of a wireless multi parameter patient monitoring system, *Procedia Technology*, 17, 542–549.
13. S. Nagender Kumar, C.M. Subas and, Sean Dieter Tebje Kelly (2015). WSN - Based smart sensors and actuator for power management in Intelligent Buildings, *IEEE/ASME Transactions on Mechatronics*, 20(2), 564–571.
14. Sherine M. Abd EI-Kader Basma, and M. Abdelmonim (2013). Smart home designs using wireless sensor networks and biometric technologies, *International Journal of Application or Innovation in Engineering and Management*, 2(3), 413–429.
15. Hanne Grindvoll, Ovidiu Vermesan, Tracey Crosbie, Roy Bahr, Nashwan Dawood, and Gian Marco Revel (2012). A wireless sensor network for intelligent building energy management based on multi communication standards – A case study, *Journal of Information Technology in Construction*, 17, 43–61.
16. L. Louise, T. Mylene, D. Mathieu, and P. Glenn (2011). Tiered wireless sensor network architecture for military surveillance application. In: *SENSORCOMM 2011: The Fifth International Conference on Sensor Technologies and Applications*, 288–294.
17. G. Padmavathi, D. Shanmugapriya, and M. Kalaivani (2010). A study on vehicle detection and tracking using wireless sensor networks, *Wireless Sensor Networks*, 2(2), 173–185.

18. K. Sohraby, D. Minoli, and T. Znati (2007) *Wireless Sensor Networks: Technology, Protocols, and Applications*, NJ: Wiley Interscience: A John Wiley & Sons Publication.

19. M. MehdiAfsar, H. Mohammad, and N. Tayarani (2014). Clustering in sensor networks: A literature survey, *Journal of Network and Computer Applications*, 46, 198–226.

20. P. Arce, J. Guerri, A. Pajares, and O. Lazaro (2008). Performance evaluation of video streaming over ad hoc networks of sensors using FLAT and hierarchical routing protocols, book, *Mobile Networks and Applications* 13, 324–336.

21. F. Bokhari (2011). Energy-efficient QoS-based routing protocol for wireless sensor networks, *Parallel and Distributed Computing, Department of Computer Science, Lahore University of Management Sciences*, 70(8), 849–885.

22. C. Intanagonwiwat, R. Govindan, and D. Estrin (2000). Directed diffusion: A scalable and robust communication paradigm for sensor networks. In: *Proceedings of the 6th Annual International Conference on Mobile Computing and Networking*, Boston, MA, 56–67.

23. D. Braginsky and D. Estrin (2002). Rumor routing algorithm for sensor networks. In: *Proceedings of the First Workshop on Sensor Networks and Applications (WSNA)*, Atlanta, GA.

24. C. Schurgers and M.B. Srivastava (2001). Energy efficient routing in wireless sensor networks, in the MILCOM. In: *Proceedings of the on Communications for Network-Centric Operations: Creating the Information Force*, McLean, VA.

25. M. Chu, H. Haussecker, and F. Zhao (2002). Scalable information-driven sensor querying and routing for ad hoc heterogeneous sensor networks, *The International Journal of High Performance Computing Applications*, 16(3), 207–219.

26. S. Guo and O.W.W. Yang (2007). Energy-aware multicasting in wireless ad hoc networks: A survey and discussion, *Computer Communications*, 30(9), 2129–2148.

27. F. Ye, A. Chen, S. Lu, and L. Zhang (2001). A scalable solution to minimum cost forwarding in large sensor networks. In: *Proceedings of the 10th International Conference on Computer Communications and Networks (ICCCN '01)*, Scottsdale, Ariz, 304–309.

28. N. Sadagopan et al. (2003). The ACQUIRE mechanism for efficient querying in sensor networks. In: *The Proceedings of the First International Workshop on Sensor Network Protocol and Applications*, Anchorage, Alaska.

29. K. Sohrabi, J. Gao, V. Ailawadhi, and G. Pottie (1999). Protocols for self-organization of a wireless sensor network, *IEEE Personal Communications*, 7(5), 16–27.

30. K. Akkaya and M. Younis (2005). A survey of routing protocols in wireless sensor networks. *The Elsevier Ad Hoc Network Journal*, 3(3), 325–349.

31. D. Xu and J. Gao (2011). Comparison study to hierarchical routing protocols in wireless sensor networks. In: *International Conference on Environmental Science and Information Application Technology (ESIAT)*, 595–600.

32. W. Heinzelman, A. Chandrakasan, and H. Balakrishnan (2011). Energy-efficient routing protocols for wireless microsensor networks. In: *Proceedings of the 33rd Hawaii International Conference System Sciences (HICSS)*, Maui, HI.

33. A. Manjeswar, D.P. Agrawal, and TEEN (2001). A protocol for enhanced efficiency in wireless sensor networks. In: *Proceedings of 1st International Workshop on Parallel and Distributed Computing Issues in Wireless Networks and Mobile Computing*, San Francisco, CA, 189.

34. A. Manjeshwar, D.P. Agrawal, and APTEEN (2002). A hybrid protocol for efficient routing and comprehensive information retrieval in wireless sensor networks. In: *Proceedings of the International Parallel and Distributed Processing Symposium (IPDPSí02)*, IEEE, 1530–2075.

35. S. Lindsey, C.S. Raghavendra, and PEGASIS (2002). Power Efficient gathering in sensor information systems. In: *Proceedings of the IEEE Aerospace Conference*, Big Sky, MT.
36. O. Younis and S. Fahmy (2004). HEED: A Hybrid, Energy-Efficient, Distributed clustering approach for Ad Hoc sensor networks, *IEEE Transactions on Mobile Computing*, 3(4), 366–379.
37. Naveen Sharma and Anand Nayyar (2014). A comprehensive review of cluster based energy efficient routing protocols for wireless sensor networks, *International Journal of Application or Innovation in Engineering & Management (IJAIEM)*, 3(1), 441–453.
38. L. Qing, Q. Zhu, and M. Wang (2006). Design of a distributed energy-efficient clustering algorithm for heterogeneous wireless sensor networks, *Computer Communications*, 29(12), 2230–2237.
39. P. Ding, J. Holliday, and A. Celik (2005). Distributed energy efficient hierarchical clustering for wireless sensor networks. In: *Proceedings of the IEEE International Conference on Distributed Computing in Sensor Systems(DCOSS'05)*, Marina Del Rey, CA.
40. S. Jung, Y. Han, and T. Chung (2007). The concentric clustering scheme for efficient energy consumption in the PEGASIS. *Proceedings of the 9th International Conference on Advanced Communication Technology, Gangwon-do, Korea*,12(14), 260–265.
41. Janvi A. Patel and Yask Patel (2018). The clustering techniques for wireless sensor networks: A review. In: *Proceedings of the 2nd International Conference on Inventive Communication and Computational Technologies (ICICCT 2018)*, Coimbatore, India, 147–151.
42. D. Koutsonikola, S. Das, H.Y. Charlie, and I. Stojmenovic (2010). Hierarchical geographic multicast routing for wireless sensor networks, *Wireless Networks*, 16(2), 449–466.
43. Y. Xu, J. Heidemann, and D. Estrin (2001). Geography-informed energy conservation for ad-hoc routing. In: *Proceedings of the Seventh Annual ACM/IEEE International Conference on Mobile Computing and Networking*, 70–84.
44. S. Misra and R. Kumar (2017). An analytical study of LEACH and PEGASIS protocol in wireless sensor network. In: *2017 International Conference on Innovations in Information, Embedded and Communication Systems (ICIIECS)*.
45. H. Tan and K. Ibrahim (2003). Power efficient data gathering and aggregation in wireless sensor networks, *ACM SIGMOD Record*, 32(4), 66–71.
46. S. Madden, M. Franklin, J., Hellerstein, and W. Hong (2002) *TAG: A Tiny Aggregation Service for Ad- hoc Sensor Networks*, Boston, MA: OSDI.
47. M. Lee and V.W.S. Wong (2005) An energy-aware spanning tree algorithm for data aggregation in wireless sensor networks. In: *IEEE Pac Rim 2005*, Victoria, BC, Canada.
48. L. Gatani, G. Lo Re, and M. Ortolani (2006). Robust and efficient data gathering for wireless sensor networks. In: *Proceeding of the 39th Hawaii International Conference on System Sciences*.
49. Noritaka Shigei, Hiromi Miyajima, Hiroki Morishita, and Michiharu Maeda (2009). Centralized and distributed clustering methods for energy efficient wireless sensor networks. In: *Proceedings of the International MultiConference of Engineers and Computer Scientists 2009 Vol I Imecs 2009*, Hong Kong.
50. H. Khattab, A. Al-Shaikh, and S. Al-Sharaeh (2018). Performance comparison of LEACH and LEACH-C protocols in wireless sensor networks. *Journal of ICT Research and Applications*, 12(3), 219.
51. J. Gnanambigai, N. Rengarajan, and K. Anbukkarasi (2012). Leach and its descendant protocols: A survey, *International Journal of Communication and Computer Technologies (IJCCT)*, 1(3), 15–21.

52. P. Manimala and R. Senthamil Selvi (2013). A survey on leach energy based routing protocol, *International Journal of Emerging Technology and Advanced Engineering (IJETAE)*, 3(12), 657–660.

53. A. Braman and G.R. Umapathi (2014). A Comparative Study on Advances in LEACH Routing Protocol for Wireless Sensor Networks: A survey, *International Journal of Advanced Research in Computer and Communication Engineering (IJARCCE)*, 3(2), 5683–5690.

54. R. Kaur, D. Sharma, and N. Kaur (2013). Comparative analysis of leach and its descendant protocols in Wireless Sensor Network, *International Journal of P2P Network Trends and Technology (IJPNTT)*, 3(1), 51–55.

55. M. Ahmad Jan and M. Khan (2013). A survey of cluster-based hierarchical routing protocols, IRACST, *International Journal of Computer Networks and Wireless Communications (IJCNWC)*, 3(2), 138–143.

56. V. Kumar, S. Jain, and S. Tiwari (2011). Energy efficient clustering algorithms in wireless sensor networks: A survey, *International Journal of Computer Science Issues (IJCSI)*, 8(5), 259–268.

57. A. Yektaparast, F.H. Nabavi, and A. Sarmast (2012). An Improvement on LEACH protocol (Cell-LEACH). In: *14th International Conference on Advanced Communication Technology (ICACT)*, 992–996.

58. N. Sindhwani and R. Vaid (2013). V Leach: AN energy efficient communication protocol for WSN, *Mechanica Confab*, 2(2), 79–84.

59. M. Chen, T. Kwon, Y. Yuan, and V.C.M. Leung (2006). Mobile agent based wireless sensor networks, *Journal of Computers*, 1(1), 14–21.

60. H. Qi and F. Wang (2001). Optimal itinerary analysis for mobile agent. In: *Ad hoc Wireless Sensor Networks, Proceedings of the IEEE*, 147–153.

61. M. Chen, T. Kwon, Y. Yuan, Y. Choi, and V.C.M. Leung (2007). Mobile agent based directed diffusion in wireless sensor networks, *EURASIP Journal on Applied Signal Processing*, 2007(1), 036871.

62. M. Chen, L.T. Yang, T. Kwon, L. Zhou, and M. Jo (2011). Itinerary planning for energy-efficient agent communications in wireless sensor networks, *IEEE Transactions on Vehicular Technology*, 60(7), 3290–3299.

63. D. Gavalas, A. Mpitziopoulos, G. Pantziou, and C. Konstantopoulos (2010). An approach for near-optimal distributed data fusion in wireless sensor networks, *Wireless Networks*, 16(5), 1407–1425.

64. M. Chen, W. Cai, S. Gonzalez, and V. Leung (2010). Balanced itinerary planning for multiple mobile agents in wireless sensor networks. In: *Proceedings of the 2nd International Conf ADHOCNETS*, Victoria, BC, Canada, 416–428.

65. W. Cai, M. Chen, T. Hara, L. Shu, and T. Kwon (2011). A genetic algorithm approach to multi-agent itinerary planning in wireless sensor networks, *Mobile Network Applications*, 16(6), 782–793.

66. I. Aloui, O. Kazar, L. Kahloul, and S. Servigne (2015). A new itinerary planning approach among multiple mobile agents in wireless sensor networks (WSN) to reduce energy consumption', *International Journal Communication Network Information Security*, 7(2), 116–122.

67. Y.-Cheng Chou and M. Nakajima (2017). A clonal selection algorithm for energy efficient mobile agent itinerary planning in wireless sensor networks. In: *Mobile Networks and Applications*, New York, NY: Springer, 1–14.

68. P.R. Vamsi and K. Kant (2014). Systematic design of trust management systems for wireless sensor networks: A review. In: *Fourth International Conference on Advanced Computing & Communication Technologies (ACCT)*, Rohtak, India, 208–215.

69. H. Deng, G. Jin, K. Sun, R. Xu, M. Lyell, and J.A. Luke (2009). Trust-aware in-net-work aggregation for wireless sensor networks. In: *IEEE Global Telecommunications Conference,* GLOBECOM 2009, 1–8.
70. E. Rehman, M. Sher, S. Hussnain, A. Naqvi, B. Khan, and K. Ullah (2017) Energy efficient secure trust based clustering algorithm for mobile wireless sensor network, *Hindawi Journal of Computer Networks and Communications* 2017(4), 1–8.
71. M.E. Fissaoui, A. Beni-hssane, M. Saadi (2018). Energy Efficient and fault tolerant distributed algorithm for data aggregation in wireless sensor networks, *Journal of Ambient and Intelligence and Humanized Computing* 10, 569–578.
72. Rabindra Bista, Yong-Ki Kim, and Jae-Woo Chang (2009) A new approach for energy-balanced data aggregation in wireless sensor networks. In: *IEEE Ninth International Conference on Computer and Information Technology, China*, 2.
73. Yasir Faheem and Saadi Boudjit (2010). SN-MPR: A multi-point relay based routing protocol for wireless sensor networks, In: *IEEE/ACM International Conference On Green Computing and Communications & International Conference on Cyber, Physical and Social Computing*, 761–767.
74. Miao Zhao and Yuanyuan Yang (2010). Data gathering in wireless sensor networks with multiple mobile collectors and SDMA technique sensor networks. In: *The WCNC Proceedings*, 1–6.
75. Babar Nazir and Halabi Hasbullah (2010). Mobile sink based routing protocol (MSRP) for prolonging network lifetime in clustered wireless sensor network. In: *International Conference on Computer Applications and Industrial Electronics (ICCAIE),* Kuala Lumpur, Malaysia..
76. Chi Yang, Zhimin Yang, Kaijun Ren, and Chang Liu (2011). Transmission reduction based on order compression of compound aggregate data over wireless sensor net-works. In: *6th International Conference on Pervasive Computing and Applications (ICPCA)*, 335–342.
77. Songtao Guo and Yuanyuan Yang (2012). A distributed optimal framework for mobile data gathering with concurrent data uploading in wireless sensor networks. In: *Proceedings of the IEEE Infocom*, 1305–1313.
78. Xi Xu, Rashid Ansari, and Ashfaq Khokhar (2013). Power-efficient hierarchical data aggregation using compressive sensing in WSNs. In: *IEEE ICC - Ad-hoc and Sensor Networking Symposium*, 1769–1773.
79. M. Soltani, Michael Hempel, and Hamid Sharif (2014). Data fusion utilization for opti-mizing large-scale wireless sensor networks. In: *IEEE ICC, 2014 - Ad-hoc and Sensor Networking Symposium*, 367–372.
80. Shiliang Xiao, Baoqing Li, and Xiaobing Yuan (2015) Maximizing precision for energy-efficient data aggregation in wireless sensor networks with lossy links, *Ad Hoc Networks, Elsevier,26*, 103–113.
81. A. Awang and S. Agarwal (2015). Data aggregation using dynamic selection of aggregation points based on RSSI for wireless sensor networks, *Wireless Personal Communications*, 80(2), 611–633.
82. P. Zhong, Y.T. Li, W.R. Liu. Joint mobile data collection and wireless energy transfer in wireless rechargeable sensor networks, *Sensors*, 17(8), 1–23.
83. Runze Wan, Naixue Xiong, Qinghui Hu, Haijun Wang, and Jun Shang (2019). Similarity-aware data aggregation using fuzzy c-means approach for wireless sen-sor networks, *EURASIP Journal on Wireless Communications and Networking*, 2019(1), 59.
84. Min Chen, Laurence T. Yang, Taekyoung Kwon, Liang Zhou, and Minho Jo (2011). Itinerary planning for energy-efficient agent communications in wireless sensor net-works, *IEEE Transactions on Vehicular Technology*, 60(7), 3290–3299.

85. Min Chen, Taekyoung Kwon, Yong Yuan, and Victor Leung. Mobile agent based wireless sensor networks, *Journal of Computers*, 1(1), 14–21.

86. Min Chen, Sergio Gonzalez, and Victor C.M. Leung (2007). Applications and design issues for mobile agents in wireless sensor networks, *Wireless Communications, IEEE*, 14(6), 20–26.

87. Min Chen, Taekyoung Kwon, and Yanghee Choi (2005). Data dissemination based on mobile agent in wireless sensor networks. In: *IEEE Conference on Local Computer Networks (LCN'05)*, 527–529.

88. Chee-Yee Chong and Srikanta P. Kumar (2003). Sensor networks: Evolution, opportunities, and challenges. *Proceedings of the IEEE*, 91(8), 1247–1256.

89. Chien-Liang Fok, Gruia-Catalin Roman, and Chenyang Lu (2005). Mobile agent middleware for sensor networks: An application case study. In: *Fourth IEEE International Symposium on Information Processing in Sensor Networks*, 382–387.

90. Preeti Sethi, Dimple Juneja, and Naresh Chauhan. Exploring the feasibility of mobile agents in sensor networks in non-deterministic environments, *International Journal of Advancements in Technology(IJoAT)*, 1(2), 296–302.

91. Francisco Aiello, Fabio Luigi Bellifemine, Giancarlo Fortino, Stefano Galzarano, and Raffaele Gravina (2011). An agent-based signal processing in-node environment for real-time human activity monitoring based on wireless body sensor networks, *Engineering Applications of Artificial Intelligence*, 24(7), 1147–1161.

92. M. Chatterjee, S.K. Das, and D. Turgut (2002). WCA: A weighted clustering algorithm for mobile ad hoc networks, *Cluster Computing*, 5(2), 193–204.

93. Y. Xiuwu, Fan Feisheng Zhou Lixing, and Z. Feng (2016) WSN monitoring area partition clustering routing algorithm for energy-balanced. In: *IEEE*, 80–84.

94. O. Singh, V. Rishiwal, and M. Yadav (2016) Energy trends of routing protocols for H-WSN. In: *IEEE*.

95. M. Wu, H. Liu, and Q. Min (2016) Lifetime enhancement by cluster head evolutionary energy efficient routing model for WSN. In: *IEEE*, 545–548.

96. P. Yadav, V.K. Yadav, and S. Yadavc (2018). Distributed energy efficient clustering algorithm to optimal cluster head by using biogeography based optimization, *Materials Today: Proceedings*, 5(1), 1545–1551.

97. K. Dasgupta, K. Kalpakis, and P. Namjoshi (2003). An efficient clustering-based heuristic for data gathering and aggregation in sensor network. In: *Proceedings of the IEEE Wireless Communications and Networking Conference (WCNC, 2003)*, New Orleans, LA.

98. P.K. Biswas, H. Qi, and Y. Xu (2008). Mobile-agent-based collaborative sensor fusion, *Information Fusion*, 9(3), 399–411.

99. P. Zhong and F. Ruan (2018).An energy efficient multiple mobile sinks based routing algorithm for wireless sensor networks in, *IOP Conference Series: Materials Science and Engineering*, 323, 012029.

100. Guoliang Xing, Tian Wang, Zhihui Xie, and Weijia Jia (2008). Rendezvous planning in wireless sensor networks with mobile elements, *Mobile Computing, IEEE Transactions On*, 7(12), 1430–1443.

101. Huan Zhao, Songtao Guo, Xiaojian Wang, and Fei Wang (2015). Energy-efficient topology control algorithm for maximizing network lifetime in wireless sensor networks with mobile sink, *Applied Soft Computing*, 34(C), 539–550.

102. A. Muthu Krishnan, and P. Ganesh Kumar (2015). An effective clustering approach with data aggregation using multiple mobile sinks for heterogeneous WSN, *Wireless Personal Communications* 90(2), 1–12.

103. Deepa V. Jose and G. Sadashivappa (2015). A novel scheme for energy enhancement in wireless sensor networks. In: *Computation of Power, Energy Information and Communication (ICCPEIC), International Conference*, 0104–0109.

104. C. Konstantopoulos, A. Mpitziopoulos, D. Gavalas, G. Pantziou (2010). Effective determination of mobile agent itineraries for data aggregation on sensor networks, *IEEE Transactions of Knowledge and Data Engineering*, 22(12), 1679–1693.

105. H. Deng, G. Jin, K. Sun, R. Xu, M. Lyell, and J.A. Luke (2009). Trust-aware in-network aggregation for wireless sensor networks. In: *IEEE Global Telecommunications Conference*, GLOBECOM 2009, 1–8.

106. E. Rehman, M. Sher, S. Hussnain, A. Naqvi, B. Khan, and K. Ullah (2017). Energy efficient secure trust based clustering algorithm for mobile wireless sensor network, *Hindawi Journal of Computer Networks and Communications*. doi:10.1155/2017/1630673.

107. Rui Wang, Zhiyong Zhang, Zhiwei Zhang, and Zhiping Jia (2018) ETMRM: An energy-efficient trust management and routing mechanism for SDWSNs, *Computer Networks Elsevier*, 139, 119–135.

108. Bibudhendu Pati, Joy Lal Sarkar, and Chhabi Rani Panigrahi (2017). ECS: An energy-efficient approach to select cluster-head in wireless sensor networks, *Arabian Journal for Science and Engineering, Springer* 42(2), 669–676.

109. Fouad Hajji, Cherkaoui Leghris, and Khadija Douzi (2018). Adaptive routing protocol for Lifetime maximization in multi-constraint wireless sensor networks, *Journal of Communications and Information Networks*, 3(1), 67–83.

110. B.M. Khan, R. Bilal, and R. Young (2017). Fuzzy-TOPSIS based cluster head selection in mobile sensor network, *Journal of Electrical Systems and Information Technology*. doi:10.1016/j.jesit.2016.12.004.

111. Govind P. Gupta (2018). Improved cuckoo search-based clustering protocol for wireless sensor networks, *Procedia Computer Science*, 125, 234–240.

112. K. Lingaraj, Rajashree V. Biradar, and V.C. Patil. (2018). Eagilla: An enhanced mobile agent middleware for wireless sensor networks, *Alexandria Engineering Journal, Elsevier*, 57(3), 1197–1204.

113. S. Sasirekha and S. Swamynathan (2017). Cluster-chain mobile agent routing algorithm for efficient data aggregation in wireless sensor network, *Journal of Communications and Networks*, 19(4), 392–401.

114. Chien-Fu Cheng and Yu Chao-Fu (2018). Mobile data gathering with bounded relay in wireless sensor networks, *IEEE Internet of Things Journal*, 5(5), 3891–3907.

115. Mohamed Elshrkawey, Samiha M. Elsherif, and M. Elsayed Wahed (2018). An enhancement approach for reducing the energy consumption in wireless sensor networks, *Journal of King Saud University - Computer and Information Sciences*, 30(2), 259–267.

116. Natasha Ramluckun and Vandana Bassoo (2018). Energy-efficient chain-cluster based intelligent routing technique for wireless sensor networks, *Applied Computing and Informatics*. doi: 10.1016/j.aci.2018.02.004.

3 Intelligent Transport Systems and Traffic Management

Pranav Arora and Deepak Kumar Sharma

CONTENTS

3.1 INTRODUCTION

3.1.1 BASIC OVERVIEW OF INTERNAL TRANSPORT SYSTEMS

An intelligent transit/transport system (ITS) is a sophisticated system, the major objective of which is to produce innovative services with reference to completely alternative modes of transportation and management of traffic, and placing users higher up in the order of priority, and thus creating a safer, more coordinated, and more efficient use of the existing transport networks. These intelligent or smart transportation systems also differ in the techniques that are used, from a simple organizational system, like basic automotive navigation, to traffic light management systems, instrumental management systems, variable message signs, automatic recognition of number plates, or speed detection cameras to observe vehicles, with a system like security cameras, in the form of various CCTV systems, as well as the use of additional complex techniques that integrate live information and review from a variety of different sources, such as guided parking systems, live weather information, and reports of accidents nearby. One such system is Google Maps which guides us to our destination, showing us the correct route, and the predicted/expected time it will take for us to reach our destination.

Figure 3.1 shows a sample of an intelligent transport system, where all the components, such as the traffic lights and the car, communicate with one another, so as to make driving possible, with less congestion and far greater safety.

3.1.2 HISTORY AND DEVELOPMENT OF THE SYSTEM

ITS systems began with the surveillance of the roads, to prevent accidents from happening and to promote a free flow of traffic. Earlier systems were funded by Governments and their agencies but, with the automation/machine learning age on its way, many private firms have stepped into this field in order to compete with each other and to create a sophisticated system. In addition to helping with the regular commute, ITS will also help in the mass evacuation of the population from localities in case of a major disaster, such as floods or a volcanic eruption.

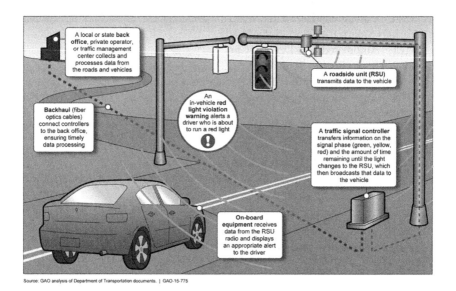

Source: GAO analysis of Department of Transportation documents. | GAO-15-775

FIGURE 3.1   An intelligent transport system

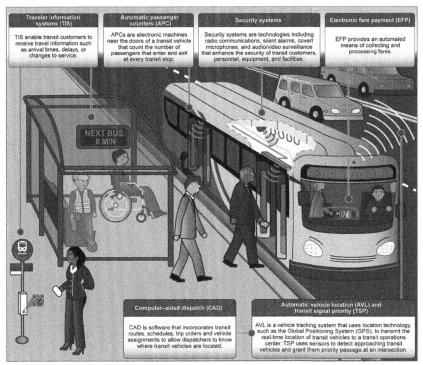

Source: GAO analysis of Department of Transportation documents. | GAO-16-638

FIGURE 3.2 An automated system for bus passengers

However, one of the major advantages of ITS will still be tackling the problem of traffic. As the population continues to grow, we will see more people wanting to commute for jobs daily, for medical needs, etc. so we must also develop a good and efficient public transport system. The high-density population areas can also benefit from such a multimodal system, where the population commutes *via* walking, bicycles, motorcycles, buses, and trains, which will also help us to fight in this never-ending battle against pollution.

Figure 3.2 shows a system where the passengers can board the bus using just a quick response (QR) code, and the bus also communicates with other nearby vehicles for them to slow down as there may be pedestrians. Developing such a system could save us precious lives and also reduce human effort.

3.2 BRIEF INTRODUCTION TO ARTIFICIAL INTELLIGENCE AND COMPUTER VISION

3.2.1 NEURAL NETWORKS

Neural networks (NN) are computer systems that are similar to and work like the biological neural networks, that constitute the brain of living individuals. Figure 3.3(a)

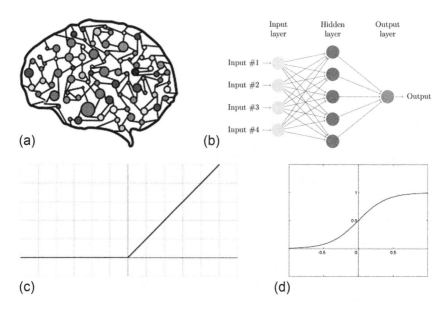

FIGURE 3.3   (a) Neural network, (b) structure layer neural network, (c) mathematical plot, (d) mathematical plot

shows a schematic way in which the neurons are present in a human brain. These kinds of system have the ability to learn and do certain tasks by taking examples, without being explicitly or specifically programmed with specific rules for the opera-tion or the task. For example, in image recognition systems, the system will learn to identify and sort images that contain dogs by analyzing and processing the example images that have been manually categorized and segregated as "dog" or "no-dog" and fed in as input. By using the results, it can then be used to identify the presence of dogs in other images. They do this without any prior knowledge of dogs; instead, they automatically generate identifying characteristics from the examples that they process, such as identifying eyes, tails, etc. A neural network consists of multiple nodes that are connected to one other and which process independently but take information from the preceding node and pass it on to the succeeding node in the layer. Such a system, consisting of multiple layers of neural networks, is called a convolutional neural network (CNN). These kinds of neural networks have major applications in image classification, image and video recognition, natural language processing, and recommender systems, as well as in various medical image analysis systems.

Figure 3.3(a) shows a neural network. Figure 3.3(b) represents a three-structure layer neural network, where the input is fed to four initial inputs or starting nodes in the input layer and they process the information, which is then passed to five nodes in the hidden (also sometimes referred as the interior) layer and then finally passed to the layer or the node, in this case, that can be used to obtain the desired result. A simple example of a neural network is one that involves the use of logistic regression, where the input parameters are tuned with the help of an activation function, like RELU or SIGMOID activation functions. We will not go into depth about what and how to use these activation functions. You need only understand the mathematical meaning of the functions, where RELU means $y = \max(0, x)$ and SIGMOID means $1/(1+e^x)$, which can be depicted in Figure 3.3, where Figure 3.3(c) represents and Figure 3.3(d) shows the mathematical plot of the respective equations.

3.2.2 MACHINE LEARNING

Machine learning is a way of achieving artificial intelligence (AI), which provides the system with the trait to automatically learn, improve, and rectify from past experience without being specifically programmed or instructed for it. The major focus is to let the machine learn automatically without any human help or intervention, and to adjust its actions with the help of the gathered data. Machine learning algorithms create a mathematical model that is based on sample input data, also known as training data, in order to make predictions or decisions without being explicitly programmed for that purpose. These algorithms rely on the patterns in the input data for the prediction or the output. These types of machine learning algorithms are used in a wide variety of applications, such as computer vision or filtering of emails, where it is not feasible or is too difficult to design a straightforward conventional style algorithm to perform the desired task effectively. As in conventional computer science, some algorithms are better at solving certain sets of problems; the same is the case with machine learning algorithms, where one performs better than the others for a set of problems.

Some major or commonly used machine learning (ML) algorithms are:

3.2.2.1 Supervised Machine Learning Algorithms

These types of ML algorithms can be applied from what has been learned in the past to a new, similar dataset, using labeled examples, to predict the outcome. The algorithm starts with the analysis of a known training dataset and then the learning algorithm produces a function that makes predictions about the output values. We will study more about supervised learning in the coming sections.

3.2.2.2 Unsupervised Machine Learning Algorithms

These types of algorithms are often used when the information that is used to train the model is neither classified nor labeled. This type of system does not predict the correct output, but, instead, it explores the datasets and draws inferences, based on the datasets, to find and depict the hidden structures in an unlabeled dataset.

3.2.2.3 Reinforcement Machine Learning Algorithms

This type of algorithm learns by interacting with its environment and, as a result, it performs some actions and obtains the error. This type of algorithm allows the machines to automatically find or obtain the ideal characteristic with respect to a specific context in order to improve the result. We do not need much information about this type of algorithm, hence it will not be discussed it further. For more in-depth details, the reader can refer to Ref. [3].

3.2.3 Supervised Machine Learning Algorithms

The definition of supervised machine learning algorithms has already been given in Section 3.2.2.1.

"Supervised" learning can be defined as a type of learning wherein the model is primarily trained on a predefined and detailed labeled dataset. When the model is provided with both input parameters and the corresponding output parameters, it is called a labeled dataset. In supervised learning, both the training dataset and the validation datasets (a validation dataset is also known as a test dataset) are labeled, as is illustrated in Figure 3.4(a).

Figure 3.4(a) has a well-labeled dataset:

   i. This shows a dataset from a retail store, that is helpful at estimating and calculating whether a particular customer will buy a certain item that is under consideration, taking into account his/her gender, age, and salary.
     Input: Gender, age, salary of the customer.
     Result: Whether purchased or not is represented by numbers 0 and 1; 1 means the customer will buy the product and 0 means that the customer will not buy the product.
  ii. It is a meteorological dataset from a random location, which will serve the purpose of predicting the speed of the wind, based on other related parameters and indicators.
     Input: Temperature and pressure at the instant, dew point, relative humidity of the air, direction of the wind.
     Result: Speed of the wind.

3.2.3.1 Training the System

When the model is trained, the pre-acquired dataset is divided according to the ratio of 80:20, where 80% is utilized in the training data and the remaining 20% of the dataset is used to test the accuracy of the trained model. In the training data, we provide the model with both input and output. The model created gains perspective and derives logic only from the training data. By training, the model will try to build some kind of logic or thinking by itself and will then predict the output for the 20% which is provided as the test data.

Once the model is prepared and trained, it is then available for testing. When testing the model, the remaining 20% of the dataset is used as the input which the model has never encountered before, and the output is then compared against the actual output from the dataset which can then predict the accuracy of our model.

3.2.3.2 Types of Supervised Learning

Figure 3.4(b) shows the types of supervised learning:

   1. **Classification**: A type of supervised learning where the result has a clear and defined value. For example, in Figure 3.4 (b), the result is that "bought" has a defined value i.e. (0 or 1), where 1 means it will be bought and 0 means it will not be bought. The aim or focus here is to estimate that the

User ID	Gender	Age	Salary	Purchased	Temperature	Pressure	Relative Humidity	Wind Direction	Wind Speed
15624510	Male	19	19000	0	10.69261758	986.882019	54.19337313	195.7150879	3.278597116
15810944	Male	35	20000	1	13.59184184	987.8729248	48.0648859	189.2951202	2.909167767
15668575	Female	26	43000	0	17.70494885	988.1119385	39.11965597	192.923834	2.973036289
15603246	Female	27	57000	0	20.95430404	987.8500366	30.66273218	202.0752869	2.965289593
15804002	Male	19	76000	1	22.9278274	987.2833862	26.06723423	210.6589203	2.798230886
15728773	Male	27	58000	1	24.04233986	986.2907104	23.46918024	221.1188507	2.627005816
15598044	Female	27	84000	0	24.41475295	985.2338867	22.25082295	233.791987	2.448749781
15694829	Female	32	150000	1	23.93361956	984.8914795	22.35178837	244.3504333	2.454271793
15600575	Male	25	33000	1	22.68800023	984.8461304	23.7538641	253.0864716	2.418341875
15727311	Female	35	65000	0	20.56425726	984.8380737	27.07867944	264.5071106	2.318677425
15570769	Female	26	80000	0	17.76400389	985.4262085	33.54900114	280.7827454	2.343950987
15606274	Female	26	52000	0	11.25680746	988.9386597	53.74139903	68.15406036	1.650191426
15746139	Male	20	86000	1	14.37810685	989.6819458	40.70884681	72.62069702	1.553469896
15704987	Male	32	18000	0	18.45114201	990.2960205	30.85038484	71.70604706	1.005017161
15628972	Male	18	82000	0	22.54895853	989.9562988	22.81738811	44.66042709	0.264133632
15697686	Male	29	80000	0	24.23155922	988.796875	19.74790765	318.3214111	0.329656571
15733883	Male	47	25000	1					

Figure A: CLASSIFICATION

Figure B: REGRESSION

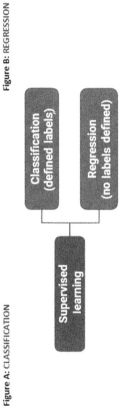

FIGURE 3.4   (a) Tabular representation of the data, (b) types of supervised learning

defined value belongs to a specific group and then the system is graded on the basis of its accuracy. Classification can also be of many types, such as a linear classification (also known as a binary classification) or a multi-class classification. In a linear classification, a model is trained to predict either 0 or 1, i.e., a binary value that is "yes" or "no" but, in the case of a multi-class classification, the model is trained to predict more than one class.

2. **Regression:** A type of supervised learning, where the output has a continuous value, say, a range or an ongoing number type of result. In the example in (Figure 3.5), the result is that the speed of the wind does not have any discrete or fixed value, but is continuous within a particular range. The main focus here is to find and calculate a value that is as close to the actual output value as our model can be and then evaluate the model by calculating the error value of the model. The smaller the error value of the model, the greater the precision of our regression model and thus the better the model. Figure 3.6 tells us, in a concise manner, about the different types of supervised learning [4].

3.2.4 COMPUTER VISION

Computer vision is defined as the study that aims to develop ways to help computers with the ability to visualize and understand digital images in the form of photographs and videos. In terms of engineering, computer vision is the ability to automate tasks that a living visualizing system is able to perform. The main aim of a computer vision system is to be able to extract information from images. Computer vision is basically a subset of artificial intelligence, where we apply various kinds of machine learning algorithms to classify images in order to maximize the accuracy

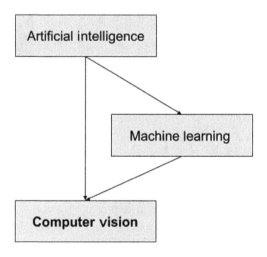

FIGURE 3.5 An ideal transportation model

FIGURE 3.6   Different signals in traffic lights

of our output in the form of a prediction. We see many examples of computer vision in our daily commuting lives, from the automated cruise control system in cars to automated parking and entry systems. Indeed, the field of computer science has revolutionized our ways to commute and will continue to do so in the future as well. Figure 3.7 shows us how computer vision is related to better-known terms like machine learning and artificial intelligence.

3.3   VARIOUS SMART TRANSPORTATION MODELS

3.3.1   AUTOMATED SELF-DRIVING CARS

A self-driving or autonomous car is a vehicle that is capable of sensing its nearby environment and moving on the road in a safe manner with no or negligible human input. Self-driving cars use various kinds of sensors to see or predict their surroundings, such as radar, GPS, sonar, odometry, speedometry, and various other inertial measuring devices. These vehicles are fitted with powerful computers, along with highly sophisticated AI algorithms, consisting of various ML algorithms and applied use of computer vision. The vehicles feed in the information, with the help of various sensors, and processes, in real time, all the scenarios that could possibly occur. The system calculates the speed of the nearby cars, as well as keeping a look-out for pedestrians, traffic signals, and road markings to create and follow a safe route. All control of the vehicle, from acceleration to braking and lane-keeping to turning is handled by the system. Driving safety experts and statisticians predict that once this self-driving car technology has become fully developed, accidents or collisions caused by human error, such as delayed reaction time caused by distraction or aggressive driving styles, will be significantly reduced. In addition to being safe,

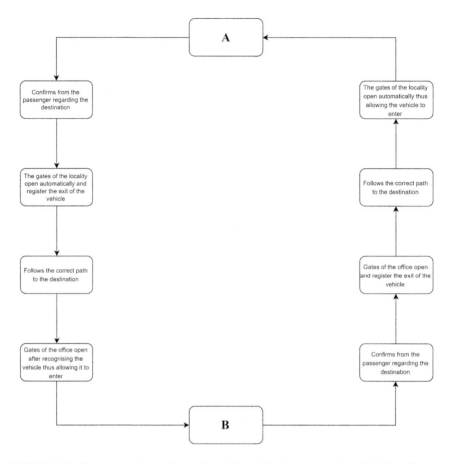

FIGURE 3.7 Computer vision relationship with machine learning and artificial intelligence

it will also allow us to travel at a much faster pace, thus reducing the daily commute times and hence making our lives more efficient. It will also provide access for people with disabilities who cannot themselves drive. Figure 3.8 depicts such a system where cars can communicate with one another.

Various levels of automation can operate in vehicles:

Level 0: The computer system has zero or almost no control over the vehicle.

Level 1: The computer system has the ability to maintain the speed of the car, but the driver must be in full control of the steering, braking and other aspects.

Level 2: The computer system has the ability to steer the vehicle at low speeds and can also accelerate and brake, when required. The driver must pay complete attention in this case also and should be in a position to take over as and when prompted by the system.

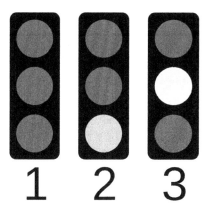

FIGURE 3.8 Automated self-driving cars

Level 3: The computer system, in this case, is in a semi-autonomous state, which means the system can control all the aspects of the car up to a certain speed that is around 60 kmph or 37 mph and requires almost no attention from the driver until that speed is reached.

Level 4: The computer system, in this case, is in a highl autonomous state and requires no intervention from the driver other than in the case of any errors or incidents.

Level 5: This computer system is the most sophisticated type, where there is no need for the driver to intervene in any case. The car, in this case, can exist without any pedals and can perform independently. We might see this system soon, in the driverless taxis of the future.

Autonomous automation has always piqued the interest of researchers, and accomplishments in this domain have been proportional. This paper presents an exhaustive chronology of the same. We have witnessed a drastic change in autonomous vehicle technology since the 1920s, and have come a long way from when the first radio-controlled vehicles were designed. In the decades that followed, we saw autonomous electric cars powered by embedded circuits on the roads. By the 1960s, the advent of autonomous cars having similar electronic guide systems stole the limelight. The 1980s saw a major breakthrough in the technology as vision-guided autonomous vehicles were invented and they haven't lost their relevance and a similar or modified form of vision and radio-guided technologies is used to date. A multitude of semi-autonomous features introduced in modern cars such as lane-keeping, automatic braking, and adaptive cruise control have been inherited from such systems only. Extensive network guided systems in conjunction with vision-guided features are how autonomous vehicles can be envisioned.

Predictions show that the companies will be able to manufacture, successfully test and launch fully autonomous vehicles as we enter into the next decade. The motivation behind the autonomous vehicles is to set foot in an ambitious era of safe and comfortable transportation and reduced traffic accidents [3].

For more information regarding how autonomous cars came into play in the current scenario and their future, you can refer to "Autonomous Cars: Past, Present, and Future" by Keshav Bimbraw.

3.3.2 SMART TRAFFIC SIGNALS

The modern three-signal traffic light or signal was created in 1920 and the basic principles have remained the same, but the aspect that has changed is the application of AI in these systems, which, with the help of a sophisticated system, has started to revamp the conventional system. But the reality is that we are very far from achieving this goal, as only 3% of all the traffic signals in the USA are "smart", despite having received heavy funding and research help from the Silicon Valley, with the majority of the traffic lights still working on basic timer-type systems.

Smart or intelligent traffic lights are a traffic system that mixes conventional traffic lights with an associated array of sensors and AI, to show intelligence route vehicles and to direct traffic. An ideal smarter traffic signal system will benefit not only the driver but also pedestrians and cyclists, and will also help to accommodate any changes in the weather that might occur.

One such system is Scalable Urban Traffic Control (SURTRAC) that was developed by researchers from Carnegie Mellon University (CMU). The system dynamically optimizes the traffic signals, hence improving the flow of traffic, leading to shorter waiting times, reduced traffic congestion, and ultimately less pollution.

The SURTRAC (Scalable Urban Traffic Control) system merges the ideas from traffic control theory with the current advancement in the domain of multiple agent planning systems and also has quite a few vital traits that distinguish it from the others. To further enhance the ascendability and dependence/reliance of the system SURTRAC firstly functions in a complete unconsolidated controlled way; every junction of the traffic is independently allocated a green period, based flow of incoming vehicles. Secondly, SURTRAC also strives to improve urban road transportation networks with a lot of input traffic flows where the coordination of individual levels is done by processing with the estimated outflows to the various downstream neighbors, which gives these traffic intersections a more informed basis for locally balancing competing for the inflow of traffic while simultaneously also promoting the setting up even of even larger green travel corridors. Thirdly SURTRAC also operates in an instantaneous manner thus each traffic junctions recalculates its initial allotment procedure and thus re-communicates expected outflows as with a frequency close to 1 per second in rolling horizon fashion, enabling both effective operations in heavily congested signal networks and also being aware to sudden changes in traffic conditions and adapting accordingly [6].

For more details, the reader should refer to "SURTRAC: Scalable Urban Traffic Control" [4].

3.3.3 AUTOMATIC NUMBER PLATE RECOGNITION

Automatic number plate recognition (ANPR) is a type of specialized training model that uses computer vision-inspired optical character recognition technology to identify the number plates of automobiles after taking/creating a temporary photographic image of them. This technique is used by federal protection forces around the world to detect individual vehicles and is also used for parking entry in certain localities. These systems use infrared sensors, along with special cameras to take a photograph of the vehicle.

The basic algorithm for such a system works in a way that it first finds the position of the number plate on the vehicle, selects that part, and then enhances the selected portion of that image to which it applies an optical character recognition (OCR) algorithm, keeping in mind the number plate regulations of that area/country. The reading of the number of plates is then used, according to the desired need.

The algorithm can be further defined in a more technical way as:

The 4 prominent steps in automatic recognition of number plates are

(1) Preprocessing of the data
(2) Finding the position of the number plate
(3) Segmentation of individual characters
(4) Recognition of individual characters [7]

For more details on how the algorithm of the system works, the reader is referred to Ref. [5].

3.3.4 AUTOMATED TOLL SYSTEMS

Automated toll systems is a wireless system that automatically collects the toll fee that is charged to vehicles. The system is equipped with an automatic radio transponder-type device, and, as the vehicle moves past the toll reader device, a radio signal from the reader then triggers the equipped transponder device and transmits back the registration number or unique ID of the vehicle, and an electronic payment system associated with it charges the user the toll fee.

One of the major advantages of the system is that the driver does not need to stop, thus reducing traffic delays. It is also cheaper to maintain or run as it only involves a one-time investment of setting up the sensors, with no human labor being employed, thus reducing costs.

The beginnings of the system date back to 1959 when William Vickrey proposed the idea for a model for electronic tolling of the Washington Metropolitan Area. The testing of the prototypes began in the 1960s and soon more countries also began working on such a system, Italy becoming the first country to deploy the system on a large scale in 1989.

One such system is the FASTag system, an initiative of the Indian Government, to tackle the problem of ever-increasing traffic.

FASTag was introduced by the NHAI (National Highways Authority of India) to ease out and eventually replace the manual toll collection system on the National Highways.

FASTag was designed to enable cashless toll payments by using the RFID (radio frequency identification) technology. The amount for each transaction is directly withdrawn from the prepaid account linked with it.

3.3.4.1   Working of the Fastag technology

The FASTag allows the user to quickly pass through toll plazas and helps the user to avoid congested traffic lanes formed as a result of traditionally slow toll collection systems. The tag is affixed to the windscreen of the vehicle and is issued by a certified bank. They are prepaid in nature and hence need to be recharged by issuing a payment either through cheque or from among a plethora of online options such as Credit Card/Debit Card/NEFT/RTGS. To make the entire transition hassle-free, toll plazas will be equipped with dedicated lanes which will be distinctly marked, and the FASTag logo will be displayed at least 70 m before the toll plaza, so that the driver is able to steer the vehicle to the correct lane. The tag reader placed at every such lane will validate your payment and the required amount will be debited from your account, allowing you rapid and successful passage.

The FASTag has numerous benefits over the existing mechanism, a few of them being:

- **Reduces time and fuel consumption**: As the transaction is cashless, there is no need to stop at the booth and pay the toll. The tag reader at the plaza will automatically withdraw the funds from your prepaid recharged amount as and when the vehicle passes through the toll plaza.
- **Transaction alerts**: Whenever the transaction is taken from the account, the FASTag user will receive an SMS, alerting them of the same. This step eliminates the fear of misuse and enhances customer safety in this whole mechanism. This also ensures that the user is aware of his account balance and of corresponding withdrawals.
- **Online recharge**: Vouching for a complete cashless mode, the recharge for the FASTag account can also be carried out online through e-banking or Debit/Credit card.
- **Eliminating the need for cash**: Shortage of smaller denomination coins/notes during a toll payment hinders the smooth process and unnecessarily takes up time. With the FASTag system, the user does not have to worry about carrying cash around.
- **User-friendly web portal**: To regularly check their expenditure at the toll booth, the customers can view their statements by logging on to the customer portal.

3.4   AN IDEAL/PERFECT SMART TRANSPORTATION MODEL

The system incorporates aspects of all the other systems that have been mentioned before to create a seamless end- experience for the user. The basic outline of the system that is being proposed is that almost no human control is required. We shall now discuss what the system actually is.

We have to move from point A (starting place of the person, assuming it to be the home) to point B (destination of the person, assuming it to be the office). The car arrives at the starting place (origin). The person then sits in the car and the car confirms where the destination, if it is a regular destination. After an affirmative response, the car begins to move, the gates of the compound/locality open automatically on seeing the recognized vehicle approach and register that a vehicle is leaving. The vehicle proceeds in the desired direction while being aware of other vehicles and following the traffic signs, that dynamically adjust on determining the number of vehicles that approach it. The vehicle reaches the destination, where the gates automatically open to allow the vehicle inside after recognizing it, and then the vehicle drops the person at the office gate and thereafter goes to the parking lot to park itself. The same operations then happen in the reverse order at the end of the day, so as to return the user to point A from point B.

Figure 3.5 summarizes all the functions in a straightforward manner and also provides an idea about how the system is supposed to work.

3.5 RESEARCH AND PROJECT IDEAS

3.5.1 Traffic Light Detection System

One of the most basic tasks one can perform is to let the vehicle identify the traffic signal that is being displayed. The machine will identify the signal and pass on the corresponding instruction regarding that signal, like 'stop' for red, and so on (Figure 3.6).

Figure 3.6 shows the different signals with respect to the standard color positioning of the signal. As in diagram 1 the uppermost light is different from the other two, indicating that it is illuminated, so, as per the standard convention, this means that it is the red light, which means 'stop'. Similarly, in diagram 2, the lowermost light is different, indicating that it is the green light, which means 'go', whereas in diagram 3 the middle light is different, indicating that it is the orange light, which means 'proceed with caution'

We can thus create a system that helps us to identify the traffic light in any given scenario. This problem can be solved using basic computer vision techniques that you might have already learned. The dataset for the given problem can be accessed at https://www.kaggle.com/mbornoe/lisa-traffic-light-dataset/kernels or https://hci.iwr .uni-heidelberg.de/node/6132 can also be used. For more information and details, please refer to the references [7–9].

BIBLIOGRAPHY

1. Gosavi, Abhijit (2009). Reinforcement Learning: A Tutorial Survey and Recent Advances. *INFORMS Journal on Computing* 21(2), 178–192. doi:10.1287/ijoc.1080. 0305. https://www.researchgate.net/publication/220668953_Reinforcement_Learn ing_A_Tutorial_Survey_and_Recent_Advances.
2. GeeksforGeeks, M.L. Types of Learning – Supervised Learning. https://www.geeksfor geeks.org/ml-types-learning-supervised-learning/.
3. Bimbraw, Keshav. Autonomous Cars: Past, Present, and Future - A Review of the Developments in the Last Century, the Present Scenario and the Expected Future of Autonomous Vehicle Technology. In: *ICINCO 2015 - 12th International Conference on Informatics in Control, Automation, and Robotics, Proceedings*, vol. 1, pp. 191–198. doi:10.5220/0005540501910198.

4. Smith, Stephen F., Gregory J. Barlow, Xiao-Feng Xie, Zachary B. Rubinstein. SURTRAC: Scalable Urban Traffic Control. gjb@cmu.edu, https://www.ri.cmu.edu/pub_files/2013/1/13-0315.pdf.

5. Automatic Number Plate Recognition System, Amr Badr, Mohamed M. Abdelwahab, Ahmed M. Thabet, Ahmed M. Abdelsadek. http://inf.ucv.ro/~ami/index.php/ami/article/viewFile/388/351.

6. FASTag. http://www.fastag.org/.

7. Bhardwaj, K.K., A. Khanna, D.K. Sharma, A. Chhabra. (2019). Designing Energy-Efficient IoT-Based Intelligent Transport System: Need, Architecture, Characteristics, Challenges, and Applications. In: Mittal M., Tanwar S., Agarwal B., Goyal L. (eds) *Energy Conservation for IoT Devices. Studies in Systems, Decision and Control*, vol. 206, Springer, Singapore, 209–233.

8. Khanna, A., S. Arora, A. Chhabra, K.K. Bhardwaj, D.K. Sharma. (2019). Iot Architecture for Preventive Energy Conservation of Smart Buildings. In: Mittal M., Tanwar S., Agarwal B., Goyal L. (eds) *Energy Conservation for IoT Devices. Studies in Systems, Decision and Control*, vol. 206, Springer, Singapore, 179–208.

9. Singh, A., U. Sinha, D.K. Sharma, (2020). Cloud-Based IoT Architecture in Green Buildings. In: *Green Building Management and Smart Automation*, IGI Global, Pennsylvania, 164–183.

10. Dandala, T.T., V. Krishnamurthy, R. Alwan. (2017). Internet of Vehicles (Iov) for Traffic Management. In: *International Conference on Computer, Communication and Signal Processing (ICCCSP)*. doi:10.1109/icccsp.2017.7944096.

11. Environmentally Beneficial Intelligent Transportation Systems (Apr 2016). https://www.sciencedirect.com/science/article/pii/S1474667016318171.

12. Hult, R., G.R. Campos, E. Steinmetz, L. Hammarstrand, P. Falcone, H. Wymeersch, (2016). Coordination of Cooperative Autonomous Vehicles: Toward Safer and More Efficient Road Transportation. *IEEE Signal Processing Magazine* 33(6), 74–84. doi:10.1109/msp.2016.2602005.

13. Moloisane, N.R., R. Malekian, D. Capeska Bogatinoska. (2017). Wireless Machine-to-Machine Communication for Intelligent Transportation Systems: Internet of Vehicles and Vehicle to Grid. In: *2017 40th International Convention on Information and Communication Technology, Electronics and Microelectronics (MIPRO)*. HYPERLINK "http://example.com?ids=" doi:10.23919/mipro.2017.7973459.

4 Data Mining and E-banking Security

Manu Bala, Seema Baghla, and Gaurav Gupta

CONTENTS

4.1 INTRODUCTION

4.1.1 E-BANKING

"Web managing" an account empowers a client to carry out money exchanges through the bank's site. This is also called "virtual managing" or "anyplace managing" an account [1]. It resembles moving the bank to one's PC at one's preferred place and time. Most customers who select web-based banking to manage their money normally identify advantages, like paying electronic bills in instalments. The e-banking services in India can be classified as below [1]:

1. **Informational e-banking**: In informational e-banking services, the server is the administrator of the banking system, and the bank's features, and items are promoted by the administrator itself. Risk-associated web management of the accounts involves the banking server itself [1].

2. **Communicative e-banking:** The bank and its clients communicate with each other using communicative e-banking, such as email, enterprise resource planning (ERP) system, requesting account operations, and using advanced applications, such as artificial intelligence, to access their personal records [1].

3. **Transactional e-banking:** Clients can carry out transactions and exchanges with the bank, using transactional e-banking services. Transactional e-banking services include account transactional information, paying various bills and invoices, interactions with other banks for fund transfers, etc. [1].

4.1.2 SECURITY IN E-BANKING

The security of data in e-banking primarily consists of three basic aspects, namely accessibility, responsibility, and secrecy. These aspects of e-banking security issues address particular concerns, such as trust of the client being lost, resulting in the client leaving the bank [2]. The main issue related to security is to keep the customer's username and passcode safe, to avoid any phishing attack incurring a loss to the customer, safeguarding the customer's e-banking details, even during a crash of the bank server [2].

E-banking basically involves electronic management of a client's account in order to help the client make e-transactions, using the bank's website and applications, benefitting the client and allowing a boost in e-business activities. E-banking services have been characterized by different banking platforms. The following salient points list out the importance and highlights of e-banking services:

- To combine the services of banking and develop innovative services.
- To encourage clients to perform transactions and make purchases through the internet, leading to savings in time and monetary charges.
- To make almost all of the exchanges electronically, without the need to physically contact any organizations.
- To manage and administer the accounting systems of the client by using data innovation, involving using a third-party banking service [3].

The common e-banking services come under following categories [2, 3]:

- Electronic charge introductions and instalment payments.
- Fund transfers.
- Recording of client's bank transactions.
- Online purchases and bill payments.
- Loan-related applications, such as loan repayments, loan re-imbursements etc.
- Support for clients with different levels of e-banking requirements for financial transactions and expertise.
- E-money exchange services.
- E-ticket booking services.
- E-bill payment services.
- E-shopping services, etc.

E-banking basically involves the use of data innovation by banking systems to encourage the use of online financial and transaction services, to save time and money incurred for various financial transactions, with proper planning for the security and ease-of-use by the client. The system estimates the client's e-banking usage capabilities and then matches it with the e-capabilities of the banking systems. Physical approaches to banking services will then be used only by older customers or by customers who do not have the knowledge or experience to operate the e-banking system. Sometimes, it can become a major issue to determine which system best suits each customer.

Customers are rapidly switching to e-banking and are more satisfied by e-banking services rather than by physical banking [3]. The business has much greater potential for growth, due to very fast money transfer systems. ATM services are the best way for customers to withdraw the money without any interface or interruptions in service, from any location at any time [3]. E-banking normally includes services such as using ATMs, credit cards, smart cards, debit-cum-shopping cards, mobile, internet, telephone banking, and electronic fund transfer systems, such as national electronic fund transfer (NEFT), real time gross settlement (RTGS), immediate payment service (IMPS), etc.

4.1.3 ATTACKS ON E-BANKING

Hackers basically use a number of ways to break into the bank's secure framework. It is of utmost importance to focus on security against such attacks. For the design of a secure e-banking system, an understanding of the various types of attacks that can be carried out on the e-banking systems is very important and must take the highest priority to achieve success of an efficient e-banking system. The following list points out the various types of attack possible on e-banking frameworks:

a) **Port Scanners**
   Port scanners can be used by attackers to compromise the security systems of the bank's frameworks. Attackers learn about various sections of the bank's framework, and use various strategies to draw and track customer data. The programming based port scanners are used by attackers to accumulate and trap secure customer data and misuse it [3].
b) **Bundle Sniffers**
   Bundle sniffers track and trap client's personal data and transaction records by utilizing a link between the client's system and the internet. To block sniffers, a secure layer is used by the developers to provide more secure banking systems that does not allow sniffers to trap sensitive information [3].
c) **Secret Word Cracking**
   Various unscrambling procedures have been adopted by hackers, using secret word breaking. If the e-banking framework does not require good passwords, these attackers can exploit system security weaknesses. So, clients using regular text make it straightforward for a secret word cracker to access a framework and to obtain important information on the client [3]. A strong password, with different combinations of words, numerals, and special symbols, can be used to avoid or at least reduce this kind of attack.

d) **Trojans**

Some hackers use Trojans to obtain a client's personal information from banking servers and database frameworks. Through Trojans, the e-banking platform screen data, texts, data exchanges, etc. can be trapped and used for malicious transactions, leading ultimately to banking frauds [3].

e) **Server Bugs**

Server bugs are basically a virus attack on the e-banking website server. These can be found and easily fixed, based on the architecture of the security system designed to restrict the hacker. The banking framework directors must be capable of detecting and removing server bugs within the least amount of time [3].

f) **Password Cracking**

Hackers use these kinds of attack to obtain the password for the e-banking login. Some of the hash table and password decryption techniques are used, such as brute force secret key assaults. Hash tables use procedures to unscramble passcodes to disclose the whole of the client's personal login data [3]. Using this approach, the hackers use techniques to obtain the user ID and passcode for a particular e-banking website.

g) **Keystroke catching/logging**

Keystroke logging is used by fraudsters to obtain passwords. The risk of being subjected to keystroke logging is more noteworthy on PCs shared by various individuals.

4.1.4   CURRENT SECURITY SYSTEMS FOR E-BANKING

Security log audits have been used to avoid various attacks. These are also used to detect the possible attackers, so as to protect the system, control data traffic, and pursue legal modalities [3]. The security system checks the sharing of data from individuals by applying security checks and audits to avoid fraudulent, logging in, or and making financial transactions etc. [3].

The security ID in present e-banking services is of utmost importance nowadays. Each client or customer is provided with a secret user ID ("user ID") and secret key or password (involving a complex combination of small and capital letters, numerals, and special symbols, up to 20 characters long) with which to access e-banking services for various kinds of activity. Various options are given to secure accounts with high levels of security, with provisions such as biometrics, one-time passwords (OTP), or email authentication. Customers can use security check procedures, such as security questions, pass code, or even requests for special permissions for sizeable transactions. The following list describes some of the highly secured options available to e-banking customers.

- **OTP** (One-time password): The client can enrol for a high-security method by associating their mobile (cell) phone number and/or email address, or any other messaging method to the account. The client obtains a security code in the form of an OTP on the respective messaging facility and uses that code for the transaction. If the account is falsely accessed by a would-be hacker, that person will not be able to carry out any financial transaction from the account without the OTP, thus securing the account against malicious attack.

- **QRP** (Quick-response protocol): This is basically a verification framework. This framework allows the financial transaction only if it receives a secret key obtained on the mobile phone, which is camera enabled. The transaction will only be possible with the confirmation token or key received on the mobile phone.
- **Security Question**: Each client account can be enabled with a security question for both transaction as well as for recovery of the account, in situations where the user has forgotten the password. For each transaction, the user needs to answer a security question; after a maximum of two failed attempts, the account becomes temporarily suspended, avoiding any malicious attack.
- **SMS Keeping Money**: Push-and-draw SMS facility can also be enabled on the client's account, using the mobile phone number. This is known as mobile banking. "Push" messages are sent to the client's mobile number regarding any transaction by the client who merely enrols for this security feature on mobile banking by paying a small charge per quarter. "Draw" messages are used by customers to know the status of the account, such as balances, by sending SMS to a specific helpline number, as requested by the customer, from time to time.

4.2 DATA MINING

Data mining (DM) has been used to extract usable information from various datasets [1, 3]. Important aspects of data mining include data preprocessing, using cleaning and missing-value treatment of data, integration of data, transformation of data, pattern evaluation and representation of data in usable form, etc. [2]. This is basically one of the tasks in the knowledge discovery process of databases as shown in Figure 4.1.

Using clustering algorithms, similar records are generally combined together to simplify the large database maintaining multi-dimensional records. The purpose of the present work is to find out where the security problem arises and how it can be avoided [1].

4.2.1 DATA MINING TECHNIQUES [2–5]

- **Association**: This process is used to detect co-occurring sets of attributes and basic rules, managing the relationships between data and attributes. Association occurs by creating relationships between different attributes in a large customer database.
- **Clustering**: Similar records are generally combined to simplify the large database, having multi-dimensional records. These are then used to create data segments that exhibit similarity within a group of points.

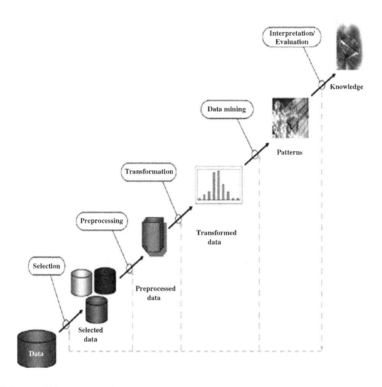

FIGURE 4.1   KDD process [1]

- **Classification**: Classification, which is a conventional DM strategy, is based on system information. In this case, attributes are classified into different categories.
- **Sequential patterns**: Sequential patterns are used to find the statistically significant patterns, where values can take on a particular order. These patterns are also related with the association rule.
- **Regression**: Regression is used to identify a relationship among different variables. Generally, this approach is based upon segmentation.
- **Segmentation**: Segmentation is the process of taking the data and dividing it up and grouping similar data together based on the chosen parameters so that it can be used more efficiently within marketing and operations. The customer bases that respond in a similar strategy are grouped together. Data segmentation gives companies the ability to identify high-opportunity groups within their customer database and can help better communication to targeted groups.
- **Prediction**: These techniques are applied for predictions, based on some sort of data or application. Regression analysis can be done for developing relationships between various variables.

4.2.2 Data Mining Applications

- **Banking/Financing**: Data mining has been used to maintain the banking and financial transaction of the client. The transactions include purchasing data, surfing data, selling data, money-exchange data, etc. [3].
- **Advertise Basket Analysis**: Data mining has also been used to record the details of the customer's purchasing information at the retailer's end. This analysis has been used by the retailers to obtain information about the customer's loyalty and to understand the specific purchasing patterns of the customer [4].
- **Producing Engineering Learning**: Data mining has also been helpful for engineering learning, to understand the connections between inventory, price, and customers, that have shown interest in a particular product [4, 5].
- **Customer Relationship Management (CRM)**: CRM is a procedure to develop a more efficient relationship and to understand the customer's behavior by gathering data and further examining the collected data. Data mining principles, such as information mining, can be used to extract the precise collection of the required information, and used for further examination, so as to improve the response of businesses [6].
- **Client Segmentation**: This is a method to classify clients, based on particular priorities, such as buying patterns, product class, typical amount spent, and method of payment. The priorities of the clients can be used to classify them and, accordingly, to target them for improving the businesses. Data mining principles has been used to find some ways to avoid clients moving to other businesses, and to impress the customers to continue offering their business by offering attractive loyalty schemes, in the form of customized offers [7].
- **Monetary Banking**: Data mining and information mining principles can be used for managing an account to record and classify specific information on the customers [7]
- **Criminal Investigation**: Data mining applications can also be used in criminology to identify wrongdoing behaviour of suspects and criminals. Data mining strategies can be applied to such arrangements, by making use of precise information in large datasets datasets, developing intricate patterns of behaviour.

4.3 RELATED WORK

Jayasree and Balan (2013) [1] described a comprehensive study on various data-mining techniques in the banking sector. The authors provided information about knowledge discovery in database processes and suggested that data mining applications and principles could be used for various applications in the banking sector. Jain and Srivastav (2013) [2] studied classification and prediction techniques for implementation in data mining algorithms, and explained the use of data mining techniques to obtain the information by descriptive partitioning of the data and by utilizing stored data, in order to build predictive models.

Ramageri (2014) [3] compared different data-mining techniques to identify their strengths and weaknesses with respect to growth in various business types. The author used various data mining techniques and algorithms to find the patterns to decide future business trends. Pakojwar and Uke (2014) [4] developed an algorithm for assessing security in online banking services, using data mining, and concluded that the banks should execute some new methods with which to verify the client's identity and to securely provide a safe e-banking platform.

Amutha (2016) [5] proposed a method to examine the information gathered from some source, with the assistance of different statistical measures like straightforward rate examination, midpoints, F-statistic, and the chi-squared test. The author suggested that statistical techniques could be used to securely improve the customer transaction analysis. Rajput (2015) [6] investigated customer's views about their experience on e-banking services, using a questionnaires with closed-ended questions and Likert types of inquiries for data collection.

Rani (2012) [7] presented a method to study perception towards e-banking in Ferozepur district of Punjab State, India, on the basis of data collected using a questionnaire. By considering the customers' opinions, it was concluded that 60% of people showed positive responses. Omariba et al. (2012) [8] presented a study on e-banking privacy and security, and examined e-banking security, and issues of protection and assaults on e-banking accounts.

Stojanovic and Krstic (2017) [9] explained the challenges and methodologies of e-banking risk management. The authors explained the 'electronic money management' method for innovatively managing the account by utilizing the principles of data innovation. The authors developed a method to determine the risks associated with e-banking. Ali et al. (2016) [10] discussed various cyber threats in e-banking services. The authors discussed various effects on customers' behavior toward e-banking, and investigated the impacts of digital dangers in e-banking web management. The authors explained how different cyber threat effects on e-banking increase awareness among the customers, and advocated the need to increase awareness among the customers when using online banking.

Shanab et al. (2015) [11] discussed various frauds and security infractions associated with e-banking. The authors explained the use of biometric data, image verification, and fingerprints for securing a bank account, and proposed that the customer must be informed and made aware of safe e-banking behavior, emphasizing money transfer risks. Devadiga et al. (2017) [12] proposed a steganography- and cryptography-based e-banking security system. The proposed method was based on storing the encrypted details in an image used to carry out the transaction. Thereafter, data mining techniques have been used to note any deviations in customers' transaction pattern or behavior.

Nwogu (2014) [13] proposed an improved e-banking security system, based on three-level security. Biometric identification-based protocols were used by the author for secure banking, who used advanced encryption standards, along with biometric identification, and concluded that biometric identification was more secure than other security tools, providing access to a more secure e-banking server rather than to customers' computers. Reddy and Reddy (2015) [14] discussed the effects of the customers' perception and satisfaction toward e-banking systems. The authors said that the use of information and communication technology (ICT) has led to improvements in the e-banking systems, and proposed that the purchaser discernment between the comfort and ease-of-use of different e-saving money administrations be measured. Consumer loyalty level toward e-banking had been distinguished.

Roozbahani *et al.* (2015) [15] explained the effect of the use of e-payment tools on e-banking on customer satisfaction. They reviewed various methods and concluded that the e-payment methods could be used to help operation of e-instalment devices and thus to improve consumer loyalty. The authors analyzed the survey data in two ways, namely (1) SPSS programming and (2) the Pearson's correlation test. Fozia (2013) [16] developed a system for secure e-management of a client's accounts, with this system using the client's transaction data and e-banking platform operational data to assess the risk of security infringement of the system in private e-banking operations. Fozia (2013) concluded that the secure e-management would help to identify the client's requirements and to ensure greater security of their account.

Jham (2016) [17] discussed the fact that banks have tried to satisfy their customers by providing the best e-services, with this paper concentrating on the utilization of internet-based savings and transactions by the bank's existing customers. The impact of the customer's acknowledgement of internet management of account administration has also been studied. Kshirsagar and Dole (2014) [18] presented a comparison of various data mining methods to identify crime detection, involving the detection of both transactional and application domains. Such detection is important, not only for governmental bodies, but also for business organizations all over the world.

4.4 METHODOLOGY

A questionnaire using a Likert scale was used to gather information on e-banking from the respondents. Basically, the Likert scale was used to measure opinions. On this scale, respondents were asked to rate their opinions on an Excel sheet. After collecting the information, the data collected were analyzed, using SPSS software.

4.4.1 ALGORITHM

- Identify the main e-banking security issues: The main factors by which e-banking security problems arise, and how, are reported.
- *Identify the influencing factors*: The major client factors determining satisfaction with e-banking security are the age, gender, and the location (urban/rural) of the client. Data on these factors were gathered from the various respondents
- *Analyze the respondents' data*: The impact of the influencing factors was analyzed, with regard to the satisfaction of the respondents with e-banking.
- *Determine the impact of e-banking platforms on respondents' satisfaction:* The impact of e-banking was statistically analyzed with respect to clients' satisfaction. Significant and highly significant values were denoted as "*" and "**", respectively.

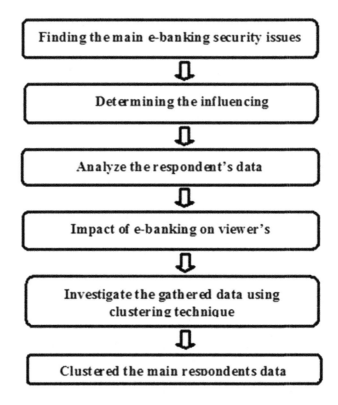

FIGURE 4.2   Flow chart of e-banking security study

* *Investigate the collected data, using clustering*: The main influencing factors were grouped using the clustering technique. The respondent's data is
  further clustered for analysis and draw meaningful inferences.

The flowchart of the work is shown in Figure 4.2.

4.4.2   QUESTIONNAIRE DATA SET

NAME _____

ROLE:
   CUSTOMERS ☐   E-CUSTOMERS ☐
AGE:
   18–30 ☐   30–60 ☐   >60☐
GENDER:
   MALE ☐   FEMALE ☐
AREA:
   RURAL ☐   URBAN ☐
TYPE:
   CUSTOMERS ☐   E-CUSTOMERS ☐

The subsequent statements relate to your attitude toward e-banking. Please point out your response to each question, on the Likert scale, of *Strongly Agree* (SA), *Agree* (A), *Neutral* (N), *Disagree* (D), *and Strongly Disagree* (SD) [5] as described in Table 4.1.

4.4.3 Implementation Using SPSS Software

After collecting the information from the questionnaires, the data collected were analyzed using the t-test, using the SPSS software tool. SPSS has been used to operate spreadsheets from MS Excel or plain text files, and relational Structure Query Language (SQL) databases as well. Figure 4.3 shows the data, relating to answers to the 20 questions (v1 to v20) on the questionnaire, from 100 customers. The columns refer to customers' data and the rows refer to the individual customers.

Data analysis view opens the data from all the questionnaires and displays the complete data in two sheets. Figure 4.4 shows menu options: Select Analyse > compare means> independent sample t-test. These options open the mean values along with statistical comparison, using the t-test.

TABLE 4.1
Likert Scale in Questionnaire

S. No.	Statement	Likert Scale				
1	E-banking provides accurate and up-to-date data	1	2	3	4	5
2	Answers individual's concerns toward e-banking	1	2	3	4	5
3	Comfortable method for managing an account and carrying out exchanges	1	2	3	4	5
4	Satisfaction with the administration through e-management account	1	2	3	4	5
5	Accurate software/hardware for serving customers	1	2	3	4	5
6	Simplicity of use for account management and cash transfers	1	2	3	4	5
7	Idea of self-administration provided by e-banking	1	2	3	4	5
8	Clarity of information on the product	1	2	3	4	5
9	Timeliness of data exchange through e-banking	1	2	3	4	5
10	Error-free exchanges through e-banking	1	2	3	4	5
11	Utilization and management of account administration	1	2	3	4	5
12	Easy tools for dealing with client's own cash	1	2	3	4	5
13	Increase in comfort and time saving	1	2	3	4	5
14	Ease of opening account or any other service	1	2	3	4	5
15	Customers can download their history of various records	1	2	3	4	5
16	Absence of proper legitimate and administrative e-banking system	1	2	3	4	5
17	Data transfer cost related to capacity/phone lines/internet	1	2	3	4	5
18	Loan facility at good rates	1	2	3	4	5
19	Unexpected system failure or framework disappointment	1	2	3	4	5
20	Security providers (cyber security, information storage, assurance of safeguard of data and protection against any fraud).	1	2	3	4	5

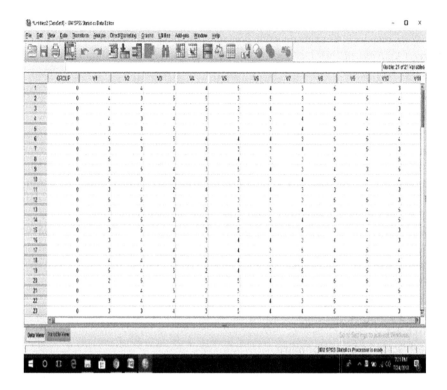

FIGURE 4.3 Variables

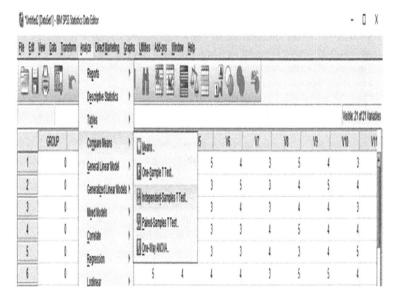

FIGURE 4.4 Data analysis

The independent t-test has been used to analyze the data, and Figure 4.5 shows how the variables are grouped together. After grouping, each of the variables is tested to determine statistical comparisons of all the values between customers and e-customers.

Figure 4.6 shows a group-statistics table and a viewer window. The system contains the menu and toolbars. The window is divided into two parts: a structured outline tree on the left and the output elements shown on the right side, i.e., a structured table of contents.

Figure 4.7 shows the output of the independent t-tests. Using the t-tests, the independent-sample means of the same continuous, dependent variable between two mutually exclusive groups (customers and e-customers) are compared.

SPSS provides some procedures for marketing applications, such as analysis, segmentation, and testing. It helps to improve interpretation of the results from the independent t-tests. Figure 4.8 shows the direct marketing option.

After carrying out the analyses, there is then a need to carry out clustering on the basis of ten inputs, as shown in Figure 4.9.

Figure 4.10 shows the data inputted given to the designated system. The interface also provides information regarding the clustering of similar data.

4.5 RESULTS AND DISCUSSIONS

The primary aim of the present work is to obtain meaningful information from the data collected, using a well-design questionnaire related to security in e-banking. Table 4.2 shows the responses of customers and e-customers, along with the t-test values.

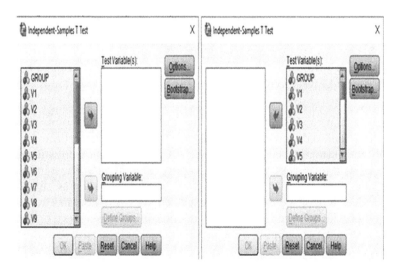

FIGURE 4.5 Grouping the variables

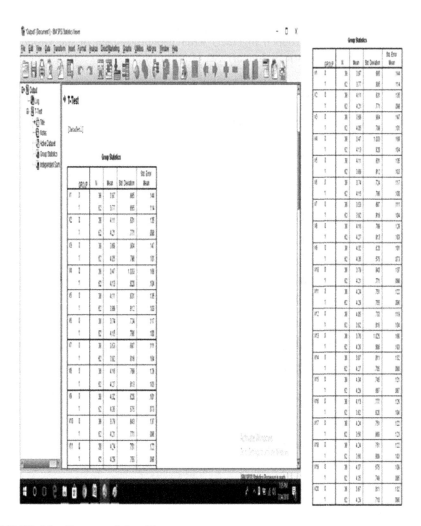

FIGURE 4.6 Group statistic table

Firstly, the data collected through the questionnaires were examined and ambiguities removed, using pre-processing of the data. The collected data are basically the customer's perception. Table 4.3 shows the ten factors found to be significant with reference to e-banking security, using the statistical independent t-test. Table 4.3 explains that the responses to questions, such as "Avoid trusting e-services to manage money", "Unexpected system failure and framework", "Security providers (cyber security, information storage, assurance of data safeguards, without any fraud)", "Not being able to maintain security", were found to be highly significant between customers and e-customers, as indicated by "**" in t-values.

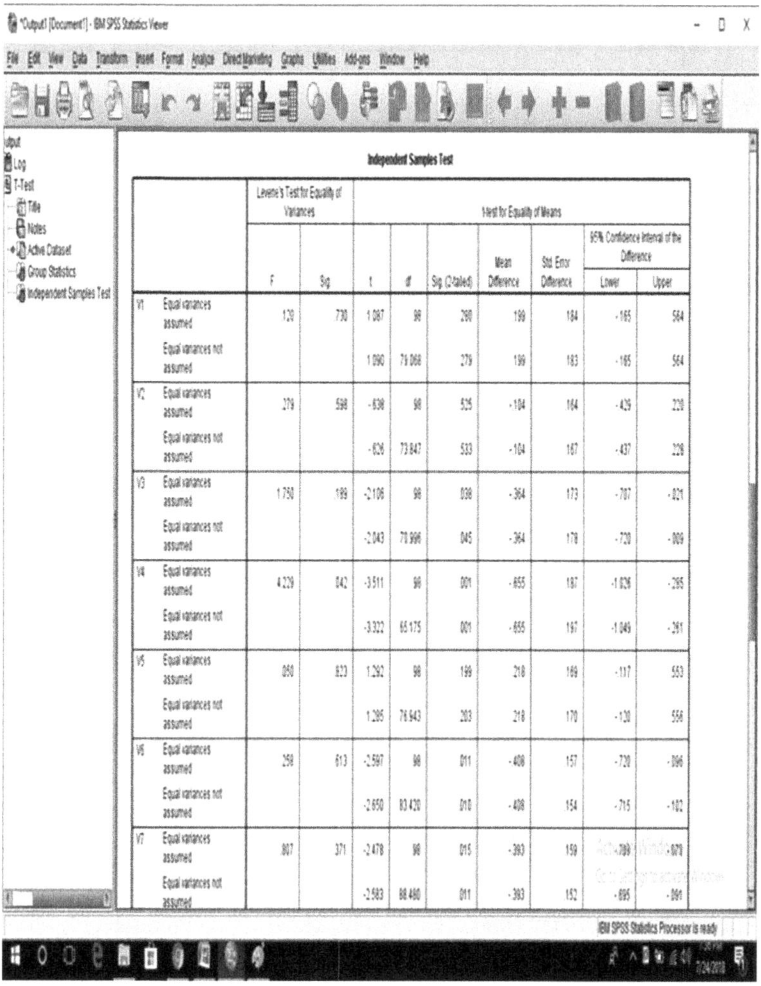

Independent Samples Test

		Levene's Test for Equality of Variances		t-test for Equality of Means					95% Confidence Interval of the Difference	
		F	Sig.	t	df	Sig. (2-tailed)	Mean Difference	Std. Error Difference	Lower	Upper
V1	Equal variances assumed	.129	.730	1.087	98	.280	.199	.184	-.165	.564
	Equal variances not assumed			1.090	79.068	.279	.199	.183	-.165	.564
V2	Equal variances assumed	.279	.598	-.638	98	.525	-.104	.164	-.429	.220
	Equal variances not assumed			-.626	73.847	.533	-.104	.167	-.437	.228
V3	Equal variances assumed	1.750	.189	-2.106	98	.038	-.364	.173	-.707	-.021
	Equal variances not assumed			-2.043	78.996	.045	-.364	.178	-.720	-.009
V4	Equal variances assumed	4.229	.042	-3.511	98	.001	-.655	.187	-1.026	-.285
	Equal variances not assumed			-3.322	65.175	.001	-.655	.197	-1.049	-.261
V5	Equal variances assumed	.050	.823	1.292	98	.199	.218	.169	-.117	.553
	Equal variances not assumed			1.285	76.943	.203	.218	.170	-.120	.556
V6	Equal variances assumed	.258	.613	-2.597	98	.011	-.408	.157	-.720	-.096
	Equal variances not assumed			-2.650	83.420	.010	-.408	.154	-.715	-.102
V7	Equal variances assumed	.807	.371	-2.478	98	.015	-.393	.159	-.708	-.078
	Equal variances not assumed			-2.583	88.480	.011	-.393	.152	-.695	-.091

FIGURE 4.7 Independent-sample test

The results for all these parameters reflected the awareness of respondents to the security issues in e-banking. They still did not have trust in the e-banking system and were afraid to use internet banking in case they lost their money. Responses to other questions, such as "Threats from hackers through encryption of login", "E-banking not being able to maintain security", "E-banking security always updated to protect any theft or fraudulent activity", "Timeliness of data through internet helps to save the account from any malicious attack. Bank quickly resolves problems customer encounters with own online transactions", "Increased comfort and time saving" were found to be significantly different between customers and e-customers, as indicated by "*" in their t-values. All these parameters reflect security issues.

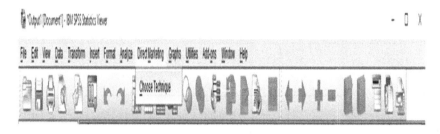

FIGURE 4.8   Direct marketing

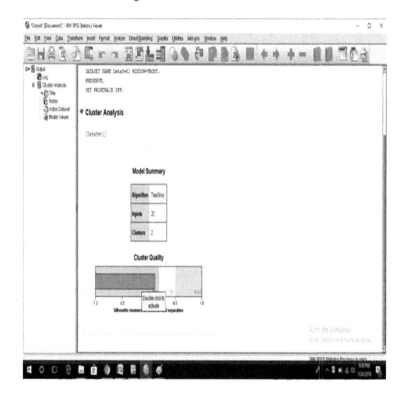

FIGURE 4.9   Cluster analysis with model summary

4.6   CONCLUSIONS

Data mining methods have been used to assess e-banking security concerns. Data from customers and e-customers were collected using a questionnaire and analyzed to identify customers' perception of security issues in e-banking. The dataset was obtained by collecting completed questionnaires filled in by 100 respondents. Clustering technique was used to find out the significant and highly significant values between customers and e-customers, using the statistical t-test. The system was analyzed using SPSS software. In this work, the customers' perception of the e-banking security has been analyzed. Ultimately, the responses were studied and

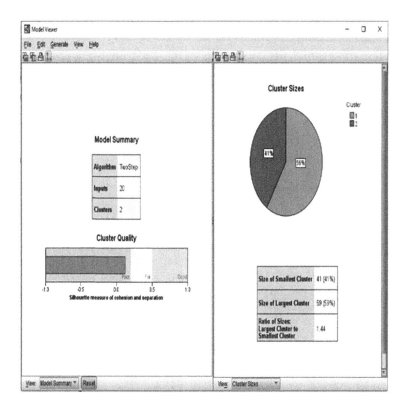

FIGURE 4.10 Cluster size

important influencing factors have been identified statistically. From the results, it has been found that e-customers are most worried about the security issues in e-banking mainly cyber security issues, information protection, fraudsters draining the login data and delayed response of the bank in case of malicious online activities.

4.7 FUTURE SCOPE

- Data from 100 respondents was collected in the present study. In future, data from web portals and other data sources can be collected to draw more meaningful conclusions from larger and more representative samples.
- Three respondent factors (age, gender, and location) were considered during the present study. In future studies, more factors should be taken into account to provide more detailed and generally applicable results.
- Using the present work, the user is only able to either like or dislike the particular facility provided by the e-banking system based on three respondent factors. In future "any suggestions" can also be applied, based on a larger number of factors.
- Further data mining software should be used, such as Rapid Miner, Orange, or KEEL.

TABLE 4.2

Perception of Respondents (1=Strongly Agree, 5=Strongly Disagree)

S. No.	Questions	Customers	E-customers	t-test (Comparing Customer and E-customer
1	Comfortable method for managing an account and transferring cash	3.97	3.77	1.08
2	Simplicity of use of e-banking to manage an account and transfer cash	4.11	4.21	0.64
3	Threats from hackers through encryption of login	3.68	4.05	2.11*
4	Lack of trust in e-services to manage money	3.47	4.13	3.51**
5	Utilization and managing an account	4.11	3.89	1.29
6	E-banking able to maintain security	3.74	4.15	2.59*
7	E-banking security always updated to protect against theft or fraud	3.53	3.92	2.47*
8	Bank has up-to-date equipment and technology	4.16	4.27	0.70
9	The customer's personal information security is better now than it was before	4.32	4.35	0.32
10	Timeliness of data through internet helps to save the account from any malicious attack	3.79	4.21	2.55*
11	Error-free exchanges through e-banking	4.24	4.29	0.34
12	Utilization and management of an account	4.05	3.92	0.82
13	Bank quickly resolves problems customer encounters with own online transactions	3.76	4.26	2.68*
14	Increase in comfort and time saving	3.87	4.27	2.64*
15	Low protection from hackers accessing the system	4.34	4.29	0.35
16	Less security during login process	4.13	3.82	1.86
17	Unexpected systems and framework failure	4.24	3.56	3.65**
18	Improved security during the e-banking usage (e.g. cyber security, information storage, assurance of safeguarding of data, and protection against any fraud).	4.24	3.66	3.55**
19	Thefts not detected immediately	4.37	4.35	0.09
20	Inability to maintain security	3.83	4.29	2.73**

TABLE 4.3
Significant Factors (1=Strongly Agree, 5=Strongly Disagree)

S. No.	Questions	Customers	E-customers	t-test
1	Threats from hackers through encryption of log-in	3.68	4.05	2.11*
2	Don't trust internet services when it comes to managing money	3.47	4.13	3.51**
3	Difficult to maintain security in e-banking	3.74	4.15	2.59*
4	E-banking security always updated to protect against theft or fraud	3.53	3.92	2.47*
5	Timeliness of data through internet helps to save the account from any malicious attack	3.79	4.21	2.55*
6	Bank quickly resolves problems customer encounters with own online transactions	3.76	4.26	2.68*
7	Increased comfort and time saving	3.87	4.27	2.64*
8	Risk of unexpected system and framework failure	4.24	3.56	3.65**
9	Improved Security using e-banking usage (e.g. cyber security, information protection, assurance of protection of client's classified data and against wholesale fraud).	4.24	3.66	3.55**
10	Inability to maintain security	3.83	4.29	2.73**

BIBLIOGRAPHY

1. V. Jayasree, V. S. Balan, "A review on data mining in banking sector", *American Journal of Applied Sciences*, 10(10), pp. 1160–1165, 2013.
2. N. Jain, V. Srivastava, "Data mining techniques: A survey paper", *International Journal of Research in Engineering and Technology*, 2(11), pp. 116–119, 2013.
3. B. M. Ramageri, "Data mining techniques and applications", *Indian Journal of Computer Science and Engineering*, 1(4), pp. 301–305, 2014.
4. S. Pakojwar, N. J. Uke, "Security in online banking services-A comparative study", *International Journal of Innovative Research in Science, Engineering and Technology*, 3(10), pp. 16850–16857, 2014.
5. D. Amutha, "A study of consumer awareness towards E-Banking", *International Journal of Economics and Management Sciences*, 5(4), pp. 350–353, 2016.
6. U. S. Rajput, "Customer perception on e-banking service", *Pacific Business Review International*, 8(4), pp. 85–94, 2015.
7. M. Rani, "A study on the customer perception towards e-banking in Ferozepur district", *International Journal of Multidisciplinary Research*, 2(1), pp. 108–118, 2012.
8. Z. B. Omariba, N. B. Masese, G. Wanyembi, "Security and privacy of electronic banking", *International Journal of Computer Science Issues*, 9(4), pp. 432–446, 2012.
9. D. Stojanovic, M. Krstic, "Modern approaches and challenges of risk management in electronic banking", *International Scientific Conference Science and Higher Education in Function of Sustainable Development*, 2(7), pp. 41–48, 2017.

10. L. Ali, F. Ali, P. Surendran, B. Thomas, "The effects of cyber threat on customer's behavior in e-banking services", *International Journal of E-Education, E-Management and E-Learning*, 7(1), pp. 70–78, 2016.
11. E. A. Shanab, S. Matalqa, "Security and fraud issues of E-Banking", *International Journal of Computer Networks and Applications*, 2(4), pp. 179–187, 2015.
12. N. Devadiga, H. Kothari, H. Jain, S. Sankhe, "E-banking security using cryptography, steganography and data mining", *International Journal of Computer and Applications*, 164(9), pp. 26–30, 2017.
13. E. R. Nwogu, "Improving the security of the internet banking system using three-level security implementation", *International Journal of Computer Science and Information Technology & Security*, 4(6), pp. 167–176, 2014.
14. D. N. V. K. Reddy, M. S. Reddy, "A study on customer's perception and satisfaction towards electronic banking in Khammam district", *IOSR Journal of Business and Management*, 17(12), pp. 20–27, 2015.
15. F. S. Roozbahani, S. N. Hojjati, R. Azad, "The role of e-payment tools and e-banking in customer satisfaction", *International Journal of Advanced Networking and Applications*, 7(2), pp. 2640–2649, 2015.
16. M. Fozia, "A comparative study of customer perception towards e-banking services provided by selected private & public sector bank in India", *International Journal of Scientific and Research Publication*, 3(9), pp. 283–292, 2013.
17. V. Jham, "Customer satisfaction with internet banking: Exploring the mediating role of trust", *Journal of Emerging Trends in Economics and Management Sciences*, 7(2), pp. 75–87, 2016.
18. A. Kshirsagar, L. Dole, "A review on data mining methods for identity crime detection", *International Journal of Electrical, Economics and Computer Systems*, 2(1), pp. 51–55, 2014.

5 Renewable Energy Sources

Kamal Kant Sharma, Akhil Gupta, and Akhil Nigam

CONTENTS

5.1 INTRODUCTION

Frequently, renewable energy (RE) is referred to as clean energy or green energy, making use of natural resources like solar orientation or rotation, or wind flow at a particular speed [1, 2]. Renewable energy is a freely available energy which is extracted from all natural resources which are present in abundance like solar orientation (which normally people frequently misinterpret as sunlight), rotation of wind (clockwise or anticlockwise direction), and using the gravitational effect of the moon towards the earth in bringing heavy tides every 15 days [3–5]. Renewable energy also involves differences in the earth's crust temperature in the uppermost layer in the form of geothermal energy, which can be connected to meet local and global loads. These sources, cited as renewable energy sources, are replicable, consistent, and readily available. Another advantage of renewable energy is its use as a dispersed resource for a connected power system.

In the past century, there was less knowledge about renewable energy sources (RES), but, as time passed, people began to understand about the generation of electricity through various RESs. With the usage of fossil fuels, like gas, coal, and oil, there are problems like hazardous gas emissions, high noise, poor reliability, and low efficiency. These non-RESs can endanger human and other life, and cause serious problems. All the RESs should be introduced by bearing many considerations in mind, because every source performs well under certain climatic conditions. Properly selecting the appropriate RESs at a particular location can provide efficient and sustainable energy. Following depletion of the ozone layer with the release of chlorofluorocarbons and halons from refrigerants and aerosols, the Kyoto Protocol was signed among developing and developed countries to save the environment by stopping the emission of harmful gases. As a consequence of this agreement, most of the countries which were not using renewable energy sources agreed to use those sources on a large scale, keeping in consideration energy demand and the type of source suitable for sustainable development.

The main reason to develop a framework for renewable energy sources is to provide immediate power and customized load ("demand") benefits. Many utilities and companies started providing in-house facilities to organizations to use RESs in a better way, to optimize their productivity and reduce carbon footprints. One of the major benefits of using RES is to provide electricity and carbon credits (which can be sold at a considerable price in the international market). Due to these benefits, many countries started promoting the usage of RES, providing different subsidies for the same. Dispersed generation, another form of energy generation, developed with changes in environmental indices, resulting in reduction of harmful gases [6, 7]. Sources of energy and their impact are qualitatively analyzed on the basis of three parameters, namely security, cost efficiency, and environmental sustainability. For over a decade, industries and communities have been working on alternative forms of energy and making them more useful in terms of localized power generation and increased capacity of alternative resources to generate electricity comparably to conventional energy sources, with transmission-distribution losses and less capital investment, resulting in a new emerging field, renewable energy sources (RES) [8].

Every emerging energy technology is based on two concepts, energy efficiency and environmental sustainability. Clean and green energy sources, in terms of sustainability, include direct or intermediate sources such as photovoltaic or wind energy (which is also a by-product of the solar system, as a result of temperature variations at the Earth's surface). Energy-efficient technologies, which build up energy efficiency, include virtual power plants, smart meters, and combined heat-power. These requirements are included in clean energy technologies [9]. Renewable energy (RE) is also dependent on hemisphere location and particular areas in terms of temperature variations. Wind generation is relatively uniform in coastal regions but, on the other hand, solar photovoltaic energy depends on solar orientation, such as sites in the southern hemisphere (or mountain-based areas). As we qualitatively analyze different sources of energy, these sources are often shown to be area specific, with investment needed changing according to selected requirements. These sectors need RE to perform under different criteria to provide efficient energy. The contribution of RE to power generation continued to expand strongly from 2012 to 2018, as depicted in Figure 5.1. Approximately 181 GW of electric power is generated in 2018 in Japan, which grows annually at 8%, making a projected total capacity of 2,378 GW globally by the end of 2018.

Greatly increased amounts of power are being generated through different RESs as compared with previous years. Globally, the hydro-power still accounts for about 60% of RE production in 2018 and is followed by wind power (22%), solar photovoltaic (SPV) (10%) and biomass power (9%). Energy sources have evolved in different aspects but, since its inception, RE has not been able to expand globally and

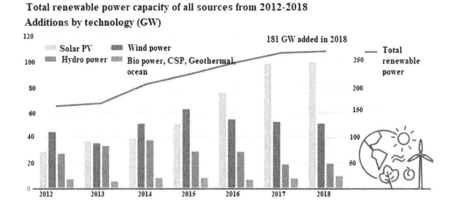

FIGURE 5.1 Growth rate of renewable energy sources from 2012 to 2018 [6]

make its presence felt in every country. The main reason for the low impact of RE globally is the presence of different temperature gradients and the location-specific nature of some RESs, although some sources are limited and can be considered only as peak sources, without 100% utilization. With under-utilization of RES, the load factor is not 100% and is unable to make inroads into into the conventional power system, resulting in limited RES usage, with particularly low impact in countries of the Asian continent.

Another major reason for the slow adoption of RES is the large capital investment undertaken in fossil fuels and correspondingly the relative lack of investment in the case of RES. Across the globe, countries are segregated into developing and developed countries. Developed countries signed the Kyoto Protocol with developing countries, to reduce carbon footprints and to reduce the risk of global warming. Despite this, developed countries, representing 35 out of 130 counties, with an installed capacity of 10 GW, are increasing their continuous productivity with the enormous use of fossil fuels, with RE in developing counties being limited to only 1 GW, as a result of rural electrification. On the other hand, in remote locations in developing counties, where access to conventional power systems is minimal and load centers are of limited capacity, implementation of RESs is easily carried out; annual increases every passing year are depicted in Figure 5.2 [10]. The European Union and islands with small capacity are converting from conventional energy to new alternative sources of energy, involving clean energy mechanisms, with solar photovoltaic (SPV) integration or microgrid (non-conventional grid), with more integrated resources connected with SPV.

5.2 CLASSIFICATION OF RENEWABLE ENERGY SOURCES

There are several RESs through which electric power can be generated, namely SPV, wind, hydro, and biomass. All these RES perform under the influence of different environmental conditions. Typical parameters of changing climatic conditions, on

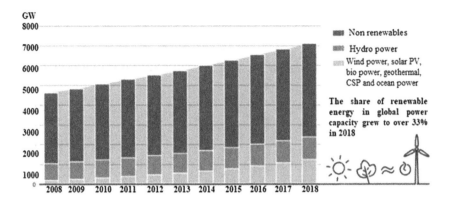

FIGURE 5.2   Global power generating capacity, since 2008 [6]

which the performance of these RES depend, include ambient temperature, wind speed, availability of run-off, solar radiation availability, and cooling conditions. This chapter highlights various avenues with which to generate electric power through available RESs, which are described below. It also reports the systems responsible for energy conversion, along with the associated technologies implemented.

5.2.1   SOLAR ENERGY

Of all the renewable energy sources available, SPV is the most tested and most abundant source available but its utilization is dependent on many parameters. It is preferred because of its fast response, high generation efficiency, and low maintenance costs. Basically, sunlight is directly converted into electricity through a non-linear semiconductor device. An SPV cell is made up of a semiconductor material, which converts sunlight into electricity. These materials are classified into three types, according to their efficiency and availability. Materials are categorized as film structures and layers of materials like single crystals, multiple crystals, uniform and non-uniform crystals with structured and non-structured arrays [11, 12]. Various crystal structures include mono-, poly-crystalline, and thin film or amorphous film used in different configurations, with different integrated circuits and voltage levels. Every installation requires different materials based on its operating function. Globally, demand for solar energy is increasing from both the consumers and the utilities.

Across the globe, the total amount of power received from the sun in the form of diffusion, emission, or dispersed power is estimated to be around 86000 TW, which could light the world without any use of conventional energy sources or the emission of harmful greenhouse gases. With advances in technology, different ways of generating electricity from solar energy have been classified into concentrated medium or solar photovoltaic medium. Energy is the by-product of each of these two processes, and the mechanism used for generating electricity arises through either the concentrated process, at a temperature of around 35°C, or *via* the solar photovoltaic (SPV) method, where generation is controlled by solar cells.

Projections are being made by different organization about the increase in generation of power by SPV to a value of 28% of global demand by 2050, as reported by the IEA (International Energy Agency) in their annual report in 2019, tabled at the annual board meeting. It is also expected in the report that the share of energy generation by SPV plants would increase from 10% to 16% by 2050, with 12% of that power being generated by concentrated power plants connected to the microgrid. Despite advances in the field of installation and implementation of modules for solar power generation, SPV module efficiency is limited to 16–25%, depending upon the type of material and its structure. Limitations in the SPV module are also associated with mono- and poly-crystalline materials, along with the direction and orientation of Sun. The SPV module also works in accordance with seasonal variations and dependence of the energy generated on the frequency of the light. Due to this dependence, semiconductor materials are used, as they behave like insulators at room temperature but, as the temperature rises and the materials have negative temperature coefficients, the semiconductor materials start to conduct. Due to this partial behavior, semiconductors like silicon or arsenide absorb photons of light and produce electrons to contribute to the flow of current. If an electron absorbs the energy of one photon, it requires more energy than its estimated work potential to release energy *via* the photoelectric effect. The SPV module also has drawbacks in the need for huge capital investment, as an installation of 1 KW is able to supply only 5 units of energy on a daily basis [7].

Solar emission and radiation are the most common methods to convert solar energy into potential sources of energy; however, solar energy is easily transformed into thermal vibrations, with the prime mover changing stabilities for conversion to electrical energy by means of a generator. The use of a generator is problematic as it involves many intermediate processes and the efficiency reduces significantly as a result, its performance depending upon the type of material used. Although many kinds of materials are available in the form of hybrid, mono- and poly-crystalline structures, the performance is still limited, and output is DC voltage, requiring a sophisticated system for conversion to AC which also introduces harmonics. Solar radiation is commonly used in SPV and incorporated in chemical and biological use as a catalyst at a specified temperature and at a particular solar orientation. SPV (solar photovoltaic) is itself a hybrid process and used for cooling and heating in remote and urban areas. It is quite environmentally friendly and economically efficient, requiring fewer installations, and certain state governments in India are offering subsidies to promote it in residential and public utilities. Especially in the northern part of India, SPV are mandatory for houses larger than 1000 sq. yards. Due to their lower efficiency, SPV systems were rarely used but research has progressed, with new materials and advanced conversion technologies. As a result, SPV has been developed as a new potential source of energy, catering for the demand of consumers, where conventional systems failed to provide the minimum electrical supply required. It is expected that, by 2030, SPV would cater for 40% of the total consumer demand in India, in areas which do not exhibit heavy demands for electricity.

There are different sizes of SPV cells available. The shape of the SPV cells may be square or round. Particular shapes are preferred, depending on the installation, as required. The SPV system consists of various components which together generate electricity. They are SPV cells, SPV module, SPV array, battery bank, charge controller, combiner box, and meters for instrumentation [13].

- **SPV Cell**: The SPV cell is a very small unit of the system. Every cell is made of a p-n junction semiconductor; in other words, it consists of a semiconductor in the form of silicon, involving the loss and addition of electrons. Every semiconductor is an insulator at room temperature, and the mobility of electrons is characterized by losing and gaining electrons in consideration of electron affinity and ionization energy. Silicon with fewer electrons but more holes and is termed as p silicon whereas silicon having more electrons is termed n silicon. A mixture of both p- and n-type silicon makes monocrystalline cells, whereas when two different materials are combined together to form a p-n junction, they are described as a non-crystalline or poly-crystalline material. Due to this combination of holes and electrons in a junction, the depletion layer is created with the energy potential, which keeps changing, depending on the biasing of a junction. For example, if a junction is forward-biased, then its width reduces, with the opposite happening in the case of a reverse-biased condition. This biasing represents the polarity of supply connected with the semiconductor materials, as the same polarity is considered to be forward biasing whereas reverse biasing is of opposite polarity, connected with the supply and junction. Biasing needs to be understood from the junction point of view, with p-type representing the positive supply (holes) and the n-type representing the negative supply (electrons) [1]. Due to this combination of junctions, I-V characteristics are determined; in the literature, V-I characteristics are also mentioned, which is the relation between load current (output current) and terminal voltage (voltage at terminals) of the cell at a specified temperature. These characteristics signify the rating of a cell at solar irradiance (solar orientation) and number of cells connected is calculated from the voltage and current behavior (Figure 5.3).
- **SPV Module**: The module represents the basic structure of a system connected together in series; it consists of many solar cells connected together, with the connected solar cells being wired and assembled together with the help of wires, and every basis wire being connected in parallel with strings. Therefore, the SPV module is a combination of solar cells with increased current rating but with every cell connected together with strings which are in parallel together. As a result, the SPV module is a series-parallel circuit combination of solar cells with the same I-V characteristics (Figure 5.4).
- **SPV Array**: As the name suggests, this is an array or collection of solar cells connected together, strengthening the current voltage characteristics. Modules consist of solar arrays connected in series to increase output terminal voltage, whereas a solar panel is connected together to form an array. An SPV array is required to generate sufficient electrical energy, with a combination of solar cells connected together, to provide a requisite output.

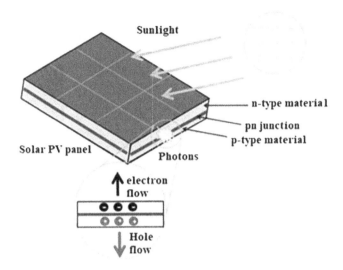

FIGURE 5.3   Schematic of a typical SPV cell

FIGURE 5.4   Schematic of a typical SPV module

In other words, for a solar cell, the solar array is an output which is further connected with conversion technologies to form a regular combination for the desired voltage and current output. SPVs are connected in a form of array for large installations, where they are connected in series or in parallel, depending upon the voltage or current requirement. For higher voltage, less load current, and less specifications of real power, series connection is preferred, whereas, on other hand, if the load requirement is greater with the minimum voltage or a constant voltage, then a parallel connection is preferred for a solar array (Figure 5.5).

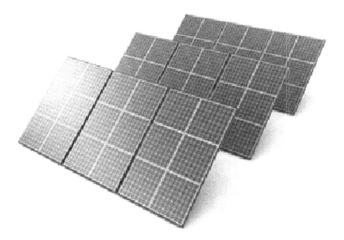

FIGURE 5.5   Schematic of a typical SPV array

- **Battery Bank**: In case of a SPV, the battery bank is an optional requirement in case the consumer wants to store excess electricity generated. This
  kind of SPV is classified into a stand-alone system or an islanded system.
  In the case of hybrid systems, where all sources are connected together,
  and the load flow analysis is carried out, then the battery bank is required
  and the SPV can be used as a peak power plant. If excess energy is not
  generated per installation, then a battery bank is not required. The main
  function of the battery bank is to supply electric power at intermediate and
  intermittent times. The main specifications of a battery bank involve discharge rates, type of material used, durability, efficiency, and ergonomics.
  A battery can also be classified into different categories, depending upon
  the system incorporated.
- **Charge Controller**: This is a microcontroller, which decodes the battery
  performance and reduces the gap between charging and discharging of the
  battery. It can be customized, depending upon the battery specification and
  the type of load connected with the SPV module.
- **Combiner Box**: This is protective equipment connected with the SPV modules and away from the installation. It is used in cases where various SPV
  modules are connected and routed with different channels. The combiner
  box involves single-conductor pigtails, with connectors already connected,
  with modules of smaller and larger capacity integrated. It also consists of a
  protection box for every string connected in parallel with the SPV modules.
- **Meter and Instrumentation**: Metering is an important aspect in SPV
  modules as they are available in off-grid and on-grid metering and available
  in two types of meters, namely the utility kilowatt-hour meter (for payment
  module) or the system meter (for calibration). These meters measure and
  display the performance of SPV systems, like battery charging and electricity used (Figure 5.6).

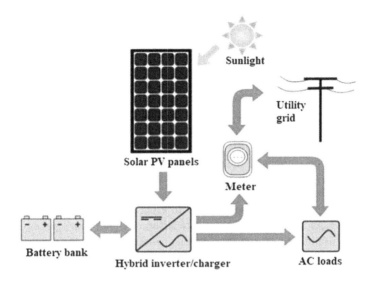

FIGURE 5.6 Schematic of a typical SPV installation

5.2.2 WIND ENERGY

Among RES, wind energy is the second most important source of electric power. This system converts wind energy into mechanical energy, which is converted into electrical energy through the use of turbines. Wind energy is attaining increasing importance worldwide, as the processes of industrialization and economic development demand improved energy production. Wind energy requires no external source; it is free. Wind energy is an intermittent source of renewable energy and therefore the power generated by wind turbines and wind farms varies, due to variation in wind speed. As a result, the performance of a wind energy system changes as a result of variation in wind speed. This system involves two types of generators which work at either constant or variable speed. Hence, the efficiency of these two types of system is different [8]. The location and installation of a wind energy conversion system is to generate the electric power near to the connected load of consumer. There are some parameters which define the performance of the wind energy system. Wind energy exceeds every other renewable energy source due to its ergonomics. As a result, wind energy is growing in importance as it is available in abundance due to temperature differences, created by solar radiation on uneven surfaces and by the elliptical rotation of the earth. It can be used at remote locations, mostly remote coastal regions, which are not supplied by conventional energy sources and is of great importance due to its contribution to the economy of a region. Wind energy is quite environmentally friendly and has a load factor less than 100%, so that it acts as a peak power plant [7].

In Asia, the term "Green Energy" has had a great impact, and most of the countries are able to use it to boost their economy and take advantage of the Kyoto Protocol, in which developing and developed countries signed an agreement to boost green energy use, giving benefits to developing countries in terms of carbon credits, valued at US\$ 14 per unit. India and China are in "rat race", developing green energy infrastructure as both are human capitals of the world and emerging economies, with huge demand for electricity. In India, rural electrification stands at 10 houses per village, remaining at 40% non-electrification. In recent years, the Ministry of New and Renewable Energy (MNRE) has been constituted in India, with similar ministries in countries throughout the world. In terms of wind energy production, India stands at 4th position globally. Some of the important facts which need to be understood before evaluating the potential of wind energy are minimum wind speed and minimum turbine height (approximately 80 m), so that the wind energy can be extracted. To date, India has been able to capture the potential in some states, with enormous wind capacity along coastal pitches. There are many hurdles to be overcome with respect to wind energy, as different types of generators are required in terms of generating electricity but due to absence of sufficient sources of reactive power compensation, control systems for pitch angles and accepting density of wind (in terms of catchment), it is sometimes difficult to control the variable output as the connected load varies.

Some of the regions in India have installed wind power plants and various companies have also invested in large numbers of turbines, due to exponential demand, but still many outcomes have been blocked due to technological roadblocks. One of the major roadblocks is interference from local governments in providing infrastructure and land, the latter being a major requirement for wind farm installations, but it is predicted that, with the formation of MNRE, states would be integrated and feasible solutions would be achieved. Considering the global scenario, major developments in wind power technologies was initiated in 2017 in India, with off-shore and on-shore technologies being established in 2018, with advances in microgrid development. With respect to on-shore wind energy generation, energy is generated in bulk and directly without any intermediate processes, but off-shore generation involves transmission of the energy generated through intermediate processes, with the efficiency reduced to 15% which is unacceptably low. In India, off-shore wind energy development potential is large but worldwide, a 4% decrease in off-shore wind energy generation has been observed in recent years (Figure 5.7).

Major factors which impact the performance of wind energy systems are:

* Variable wind speed operation of generators to generate maximum energy,
* Advances in power electronic devices for wind energy,
* Improved power plant operation and efficiency [14, 15],
* Better economics, due to the presence of large wind-generating plants.

The following are the components of a WPP (wind power plant), along with their roles and limitations:

* **Nacelle**: This is basically a combination of components connected together at the top of the wind turbine, can be seen in top-view projections of a plant, and includes wind turbine subsets. Nacelle, as the name suggests,

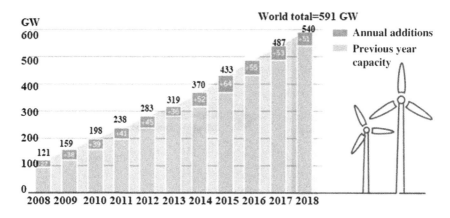

FIGURE 5.7 Wind power global capacity since 2008 [6]

connected various components of a generator, with the controlling mechanism, including output and input transitions of a different nature. It also
includes power electronic interfaces with values required to drive a generator with sufficient amount of reactive power injected into it. The main
role of the nacelle is to maintain uninterrupted motion of the wind plates
in a direction so that optimum rotation to achieve maximal power can be
obtained [11] (Figure 5.8).

- **Rotor**: The rotor picks up the wind energy and passes it on to the drivetrain.
  The rotor hub is a component which connects the rotor blades to the rest of
  the system and delivers the power to the rotor shaft. The position of each
  individual rotor blade, with respect to the wind, can be controlled using so-
  called pitch motors (Figure 5.9).

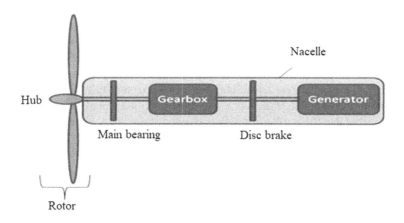

FIGURE 5.8 Nacelle of a wind power system

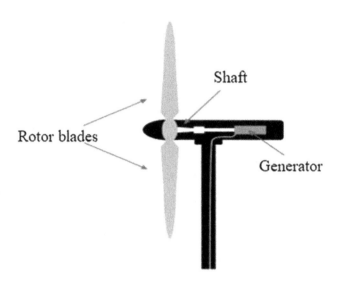

FIGURE 5.9   Rotor of a wind power system

- **Tower**: This is the largest and heaviest part of the wind turbine. The higher the tower, the lower the loads due to turbulence, resulting in higher wind speeds. There are different types of towers, such as tubular steel towers, lattice towers, and guyed pole towers.
- **Turbine**: Turbines act as a low-pressure to high-pressure mechanism, working opposite to a nozzle in such a way that its operation is similar to a prime mover which drives a generator at a particular speed to generate electricity. In a wind power plant, the input of the wind turbine is variable and wind speed is not constant, so the importance of the wind turbine increases significantly and is classified on the basis of operation and requirement. Rotation of the turbine depends on the axis of the wind, designated as horizontal or vertical, so that the turbine needs to be connected perpendicular to the axis to achieve its maximum output, with the wind turbines being designated as horizontal or vertical axis turbines Wind turbines can be divided according to the orientation of the axis of rotation, such as horizontal axis and vertical axis turbines. The two turbines are different in nature, with different terminologies, with horizontal turbines requiring a tower and a nacelle at the top, and requiring skilled manpower for installation and handling of large amounts of wind speed. These kinds of turbines are costly in nature and the number of blades is variable, depends upon the requirements and centrifugal forces dynamics. Mostly, the number of blades is limited to three, as increasing number of blades will reduce the efficiency of the system; the more complex in nature, the greater the reactive power requirement and the less the voltage control. On the other hand, vertical turbines are less costly and are simple to construct but handle very less wind and are used as off-shore wind power plant; they are limited to

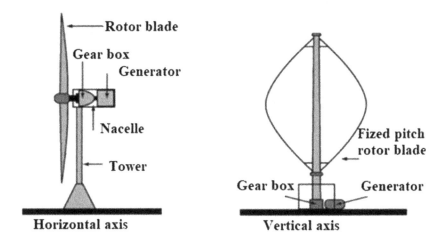

FIGURE 5.10   Classification of wind turbines

low power generation at remote places with considerably less manpower. Vertical turbines have lower conversion efficiency and are used only with lower blade effectiveness (Figure 5.10).

- **Generator**: There are different types of wind generator, the effectiveness of which depends upon the speed. They are referred to as fixed-speed induction generators or doubly fed induction generators.

  a) **Double-fed induction generator**: Double- (doubly) fed induction generators involve feeding at both ends, with excitation as an external source parameter. This type of generator is variable and can be customized; it is non-synchronous in nature as active and reactive power controls can be performed by changing the frequency of a system and by partial voltage control at the rotor side. With a change in excitation, the field changes and, by changing the field above and below the power factor, efficiency can also be affected. These operations of the generator can be studied through the V-curves of synchronous machines, with changing field excitation. As an induction generator, the operation also depends upon the value of slip; it will work only if the value of slip is negative, with reactive power flowing. This kind of generator comprises of two parts: the stator and the rotor, where the stator is connected with the grid but, in other rotor operations, is controlled by excitation and driven by changes in speed and frequency. Moreover, these generators are used primarily because of their range of speeds, which provides a range of generators with variable frequency and speed, and customized for different plants with suitable compensation (series and shunt) (Figure 5.11).

  b) **Fixed-speed induction generator**: Wind power plants are subjected to induction generators − fixed or variable. The fixed-speed induction generator follows the same operation as the doubly fixed but is of a

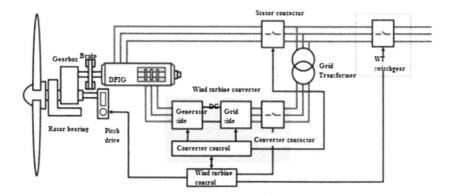

FIGURE 5.11   Schematic of a typical doubly fed induction generator

squirrel cage type and is considered to be robust. Its speeds are not controllable and it is used for low levels of power generation at a constant input of wind speed. This kind of generator is cheap and easy to install, but its efficiency and survival rate is low, due to its lack of ability to withstand fluctuations in load. Another major drawback is its inability to control reactive and real power, an external source being required for compensation, which is quite costly due to its specifications and its low availability (Figure 5.12).

5.2.3   HYDRO-POWER

Hydro-power is commonly regarded as a conventional energy source but a hydropower plant can also be classified as a non-conventional energy source, with a run-off river power plant, with or without pondage. Hydro-power plants used as renewable energy sources are not able to sustain a load factor of 100% and are categorized as peak power plants, with indices of real and reactive power. Hydro-power plants convert potential energy to kinetic energy, using different types of turbines classified on the basis of their heads. Hydro-power plants use different heads and require specific

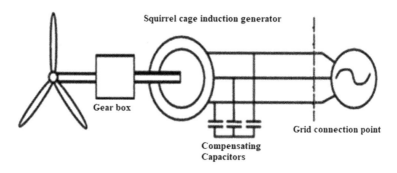

FIGURE 5.12   Schematic of a typical fixed-speed induction generator

turbines to act as a prime mover for synchronous generators. In a hydro-power plant, low-speed synchronous generators, with salient types of heads, are used in a vertical position, with the water falling from the vertical side. This kind of power plant consists of a reservoir, dam, and catchment area, and serves a large, dispersed population [11]. Different types of heads are maintained, depending upon the capacity of the generator, and to prevent the dam from excessive water, different types of gates are available and are termed spillways. Hydro-power plants cater for peak demands and base demands of generation and are monitored through real-time scaling. In India, hydro-power plants cater for 40% of total demand and globally they meet 60%, with an estimated capacity of around 1,132 GW.

Since 2017, India has increased its installed total capacity by 1% and commission of plants by 0.5%. By the end of 2018, India had increased its hydro-power capacity to a total capacity of 45.1 GW. There are advantages of hydro-power such as its being environmentally friendly, with high flexibility, and low operational and maintenance costs, whereas disadvantages are its dependence on rainfall, and the high capital costs and the remote locations of hydro-power plants. Important components of a hydro-power plant are illustrated and explained below:

- **Dam**: The dam plays an integral role in constructing a hydro-power plant with different heads, forming a reservoir and controlling the water, when needed. There are different kind of dams available, which are classified on the basis of area, construction type, and capacity to be created, known mainly as rock-fill, earth-fill, and masonry dam. Masonry dams are very few in numbers and are located only near deltas, whereas rock fill dams are robust and can have a large capacity in terms of energy generated, with earth fill dams being semi-built, with underground dams. It is essential to maintain the dam, with spillways provided to release the build-up of excessive amounts of water.
- **Water reservoir**: The reservoir is a part of the dam, which stores the water in a rainy season, providing it, whenever required, with a help of surge tank if excessive water has to be pumped out. The reservoir is the main bank, where water is to be released for making different heads, so that the height of the reservoir should be more than the height of the dam, with water held by the dam. The dam height is also used to measure available potential energy of the water stored to be used for generation of electricity.
- **Trash rack**: This is a preventive measure to remove any unwanted material from the water released from the reservoir to dam. Its main function is to prevent blockage of nozzles and turbines with such material. Most hydro-power plants with different heads use a reaction turbine, but some of the plants use an impulse turbine where the pressure on the water needs to be the same on both sides of the dam; debris can destroy an impulse turbine and also keep water in its liquid form when the temperature is low.
- **Forebay**: This is basically a regulatory mechanism, which involves a surge tank to control the flow of water, taking into account the value of the load. If the load is suddenly reduced, then generation of electricity has to be

slowed down but water intake has already been carried out; therefore water regulation is used in the form of forebay so that penstock, the channel which carries most of the water, is not destroyed.

- **Generator**: Different types of generators are used for hydro-power plants, and are termed synchronous generators; the generators need to be of the salient type to keep the transient reactance constant throughout, and used with low-speed capacity to generate electricity. They are connected in a vertical formation, having real and reactive power ratings associated with the control mechanism.

- **Penstock**: This is a channel for carrying water from the reservoir to the processing unit. In this case, the processing unit involves the generator, turbine, and intermediate processes, and the outcome is the generation of electricity. This is a shaft sometimes used to bring water from the reservoir to the processing unit through a zigzag route in which a pumped motor is required to speed up the water flow to the end of the water turbine. To prevent the need for a penstock, the surge tank is connected to maintain the overflow of water back to the reservoir, which can otherwise damage the penstock. The penstock is a wide channel, which can be opened for maintenance in earth-fill plants.

- **Water turbine**: The turbine plays an integral role in a hydro-power plant for electricity generation. The turbine is the most expensive and important component. The turbine is the heart of the system, converting potential energy into kinetic energy and acting as a prime mover. Depending upon the use of different heads, turbines are classified as impulse or reaction turbines. Impulse turbines are used with fixed heads, whereas impulse turbines are used with variable heads. In the case of reaction turbines, they are classified as Pelton, Kaplan or Francis turbines, which are selected in order to generate optimum output. Water flows from the penstock channel to the turbine and, after the designated operation, water flows down with help of the tail race so that the turbine should be free from moisture, and the blades should be mechanically strong and free from friction, and can be used for longer periods of time.

- **Transmission lines**: The main work of the transmission network is to transmit electricity to end-users, keeping in mind the protection level and ergonomics. Transmission lines can be categorized as long, medium or short transmission lines, depending upon voltage level and the length connected. Certain parameters are involved to maintain the configuration of the transmission line, following its ABCD parameters and delta star operations.

5.3 CONCLUSION

RESs are predicted to play a vital role in power production for many years. These RESs employ advanced technologies, provide faster response and achieve greater efficiency. Globally, reports have shown the total estimated capacity of all RESs worldwide. These RESs are expected to play a significant role in development of the installation of smart grids, thus paving the way for supplying good quality electric power to consumers. Obtaining good quality electric power will greatly enhance the confidence of industrial players to contribute more in building infrastructure which can greatly help in further enhancing the economy.

REFERENCES

1. C. Wang, and Yuefeng Lu, Solar Photovoltaic, Bachelor thesis, Savonia, p. 35, 2016.
2. M.M. Rahman, S. Salehin, S.S.U. Ahmed, and A.K.M. Sadrul Islam, "Environmental impact assessment of different renewable energy resources: A recent development," *Clean Energy for Sustainable Development*, pp. 29–71, 2017.
3. P.R. Varun, I.K. Bhat, and I.K. Bhat, "Life cycle greenhouse gas emissions estimation for small hydropower schemes in India," *Energy*, 44(1), pp. 498–508, 2012.
4. R. Bertani, and I. Thain, "Geothermal power generating plant CO_2 emission survey," *IGA News*, 49, pp. 1–3, 2002.
5. F. Blaabjerg, and K. Ma, "Renewable energy systems with wind power," *Power Electronics in Renewable Energy Systems and Smart Grid: Technology and Applications*, edited by B.K. Bose, John Wiley & Sons, Inc, pp. 315–345, 2019.
6. REN21, Renewable 2019: Global status report. Secretariat Renewable Energy Policy Network for the 21st Century (REN21) Paris, 2019.
7. S.M. Islam, C.V. Nayar, A.A. Siada, and M.M. Hasan, "Chapter 1-Power electronics for renewable energy sources", *Alternative Energy in Power Electronics*, Butterworth-Heinemann Elsevier, pp. 1–79, 2011.
8. P.S.R. Murty, "Chapter 24-Renewable Energy Sources", *Electrical Power Systems*, Butterworth-Heinemann Elsevier, pp. 783–800, 2017.
9. W. Strielkowski, "Renewable Energy Sources, Power Markets, and Smart Grids", *Social Impacts of Smart Grids*, pp. 97–151, 2020.
10. B.K. Sahu, "A study on global solar PV energy developments and policies with special focus on the top ten solar PV power producing countries," *Renewable and Sustainable Energy Reviews*, 43, pp. 621–634, 2015.
11. H.J. Wagner, "Introduction to wind energy systems," *EPJ Web of Conferences*, 54, p. 01011, 2013.
12. I. Dincer, "Renewable energy and sustainable development: A crucial review," *Renewable and Sustainable Energy Reviews*, 4(2), pp. 157–175, 2000.
13. E. Fertig, and J. Apt, "Economics of compressed air energy storage to integrate wind power: A case study in ERCOT," *Energy Policy*, 39(5), pp. 2330–2342, May 2011.
14. W.B. Lowrance, and H. Dehbonei, "A versatile PV array simulation tools," presented at *ISES Solar World Congress*, Adelaide, South Australia, 2001.
15. A.K. Aliyu, B. Modu, and C.W. Tan, "A review of renewable energy development in Africa: A focus in South Africa, Egypt and Nigeria," *Renewable and Sustainable Energy Reviews*, 81(2), pp. 2502–2518, January 2018.

6 A Review of the Internet of Things (IoT)
Design and Architectures

Rohit Kumar, Gaurav Bathla, and Hramanjot Kaur

CONTENTS

6.1 INTRODUCTION

The Internet of Things (IoT) has attracted much interest recently and this technology has grown and developed in leaps and bounds. IoT has penetrated deep into human life and has been immensely beneficial, with applications ranging from automobile automation to health care. The phrase "IoT" was first created by authors in 1999 [1], when Kevin Ashton was working on an intelligent network consisting of sensor devices, wireless devices, actuators, robotic arms, etc. In the modern era the term IoT can be defined with the equation:

$$IoT = Sensors + Data + Network + Services$$

According to the report and statement published by International Telecommunication Union (ITU) in 2005 [3], IoT will be a major driving force and will facilitate the construction of smart and intelligent sensory networks, connecting the globe. IoT has also been a key driving force with respect to supportive technology for pervasive (ubiquitous) computing, and the impact of IoT can be seen in mobile devices, smart watches, digital cameras, tablets, etc. [4]. The penetration of IoT has led to an immense increase in data traffic across the IoT networks, which will increase manifold in the near future; according to Cisco data, the traffic on IoT-based networks in 2019 was 70% higher than in 2014, when the growth of traffic was only 40% in the year 2009 [5]. Similarly, it is forecast that the number of connections from one device to another will increase at the rate of 43% from 2014 onwards [6]. This increase can be credited to the fact that IoT has been involved in a large number of scenarios and devices, e.g., radio-frequency identification (RFID) tags, wireless sensors, actuators, Bluetooth-enabled devices, intelligent driving systems and cars, smart homes, and many more technologies. Looking at the bright future of IoT, companies like Intel, IBM, Cisco, etc., have developed their specific reference-based architecture for IoT. Due to specific applications and vendors' motives, a single standard reference model for IoT has not been finalized [8]. In this chapter, IoT core concepts and different architectures will be presented and discussed. In Section 6.1, the core elements of the IoT network are presented in tabular form; following on from a description of IoT basics in Section 6.2, in Section 6.3, IoT-specific features are presented and, in Section 6.4, consideration of various architectures is summarized. Section 6.5 presents a comparative analysis of the different architectures, and Section 6.6 describes the main conclusions of IoT development.

6.2 BASIC ELEMENTS OF IoT

To understand the most important aspects of IoT, the reader should have a good understanding of the foundation blocks of this emerging technology. Implementation of IoT requires an understanding of various simple fundamentals, as depicted in Table 6.1. These include sensors, things or devices, actuators, networks, gateways, cloud infrastructure, and RFID tags [9].

TABLE 6.1
Fundamental Elements of IoT

Elements of IoT	Description
Sensors	Sensors are the primary means of data collection and monitoring from the given environment
Actuators	Actuators perform actions and help in performing operations and data collection without human involvement.
Gateways	They act as a bridge across different IoT components and ensure smooth information flow.
Network infrastructure (NI)	IoT is essentially a computational communication network and encompasses all major components of conventional computer networks, like routers, gateways etc.
Cloud infrastructure (CI)	IoT can be easily integrated with Cloud-based networks for better storage capability and computational capabilities.
RFID	Radio-frequency identifiers (RFID) and other bar-code-like technologies, which can help in object detection with ease.

6.3   ARCHITECTURE

This section showcases the various architectures introduced by researchers for the IoT. Many architectures exist in the literature, that have been proposed in the past by various researchers during the rise of IoT technology. These architectures can be classified as :

- Layer-specific architectures.
- Domain-specific architectures.
- Industry-defined architectures.

Further classification of the aforesaid architectures is presented in Fig. 6.1.

6.3.1   LAYER-SPECIFIC ARCHITECTURES

Categorization of IoT on the basis of layers is shown in Fig. 6.2.

6.3.1.1   3-Layer Architecture

This is the elementary design for the Internet of Things. According to [10], it includes three layers, which are explained as:

i. *Perception Layer,* also known as the sensor layer or the recognition layer [11], is the bottom layer of IoT architecture. This layer interconnects the real objects and devices *via* basic elements of IoT (explained in Table 6.1). The main functionalities of the perception layer are:
   - To connect real objects with an IoT-based system.

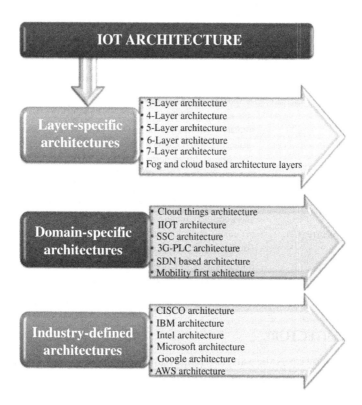

FIGURE 6.1   Categorization of IoT architecture.

- To calculate, assemble, and handle the state-of-the-art information associated with the objects *via* smart gadgets.
- To pass the data, after processing, on to the upper layer by the use of layer interfaces.

ii. *Network Layer* is the central layer of IoT architecture, and is also known as the transmission layer or gateway layer [12]. The network layer is the primarily focused layer in IoT architecture, as it basically builds the assimilation of various devices (gateway, hub, switches, or cloud infrastructure, etc.), as well as various data transmission technologies (like Bluetooth, long-term evolution (LTE), WiMax, etc.).The objectives of the network layer are:

- To obtain the processed information that is handed over by the perception layer.
- To regulate the routes for transferring this information to different IoT devices through interconnected networks (e.g., WiFi, Router, Access point, Bluetooth etc.).

iii. *Application Layer* is the uppermost layer in the IoT architecture [13]. It is also known as the business layer and performs the following major functions:

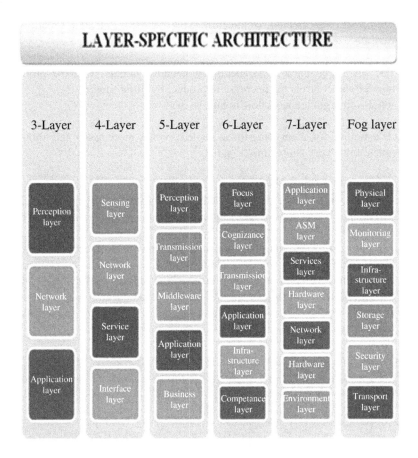

FIGURE 6.2   Classification of layer-specific architecture.

- To receive the information transmitted from the network layer.
- To use these data for distribution among the required services [10].

Regardless of the benefits, like straightforwardness, effortless problem identification, flexibility, and management functionalities, the absence of security in this layer is a crucial drawback [14]. The network layer and application layer perform complex functions, For instance, alongside finding the routes and sending the information, the network layer should also offer other data services such as data fusion and computation, etc. In addition to providing customer services, this layer must also provide information services, like information analytics and mining etc. [13, 15].

6.3.1.2   4-Layer Architecture

The author of [16] defines a layer as performing a task intermediate between the network and application layers. This service layer is held accountable for delivering services in the Internet of Things. Based on this concept, the IoT researchers have introduced a Service-oriented Architecture (SoA) [13, 15]. The main functions provided by four-layer architecture are:

- The IoT based on SOA conceals the needless details from the product builder and controls the decrease in product building time.
- It helps to organize workflow in an easy way.

It builds a method for advertising the saleable products more easily and allows it to be completed in a very short time (Fig. 6.3).

6.3.1.3   5-Layer Architecture

A five-layer IoT architecture was proposed by the authors in [18] and is depicted in Fig. 6.2. The main objectives of all the layers under this model are discussed below:

    i. *Perception layer* is mainly associated with physical objects. This layer observes and collects the data about physical objects (e.g., humidity and temperature of the environment, location and motion of the objects, etc.) and further transmits information to the next upper layer.

    ii. *Transmission layer* is used to achieve the security objective and assists in reliable transmission of information to the next upper layer.

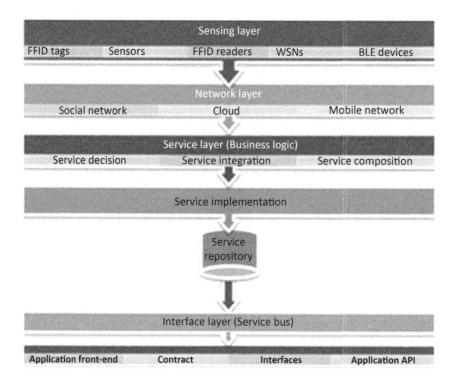

FIGURE 6.3   4-layer SOA for IoT [17].

iii. *Middleware layer* is concerned with information storage but it also processes the data and takes appropriate actions, based on the results.

iv. *Application layer* achieves the functionality of application management, such as smart grids, smart healthcare, smart home, smart farming, etc.

v. *Business layer*, the uppermost layer, supports data analytics, on the basis of which it plans further actions [19].

In [8], the five-layer architecture is described, in which the five layers are:

i. *Perception layer*, which is the bottommost layer, and it perceives the data.

ii. *Transport layer* has the same functions as the network layer in the three-layer model, i.e. it transmits the perceived information from perception layer (or from devices) to the application layer.

iii. *Processing layer* stores, analyzes, and processes the data to identify some definite pattern which can be exploited for use by the business layer, which manages the entire IoT.

iv. *Application layer* manages the applications.

v. *Business layer* carries out the data analysis.

6.3.1.4 6-Layer Architecture

A six-layer architecture was introduced in [20], based on the earlier-defined 4-layer model, but possessing two new layers. These are:

i. *Media access control (MAC) layer*, which assists in observing and controlling gadgets. It also helps in saving the energy of the devices by allowing them to sleep in their idle time.

ii. *Processing and storage layer*, which assists by carrying out the queries, analyzing and storing the information and providing security features[19].

A six-layered architecture for IoT was proposed by the authors in [21] is shown in Fig. 6.2, Roles of each layer are described as follows:

i. *Focus layer* is mainly focused on the recognition of devices, and it takes account of the multiple facets of the system under observation.

ii. *Cognizance layer* determines the sensing competence (for this purpose, sensors or actuators are deployed) and it gathers data from the physical objects that are observed in the previous layer.

iii. *Transmission layer* allows the sending of perceived information received from the cognizance layer to the subsequent layer.

iv. *Application layer* achieves the collection and classification of the data according to the requirement of the desired zone.

v. *Infrastructure layer* assists with the features which are linked with service-oriented architecture, such as cloud and cognitive computing, geographic information system (GIS) mapping, big data, and data storage functionality.

   vi. *Competence business layer* works to scrutinize the commercial networks of the Internet of Things systems. It also provides privacy and this layer is also concerned with the business models, as well as profit models, in an efficient manner.

6.3.1.5   7-Layer Architecture

By considering the physical objects or devices in the surrounding environment, researchers projected a seven-layer architecture [22] as shown in Fig. 6.2. This enhanced architecture has some special features. The users can consider different aspects that influence the sensing potential of actuators and sensors, etc. Various roles and functionalities of these layers are described [21]:

    i. *Application layer* gathers the data about a specific task, based on the client's requirements.
   ii. *Application support and management layer* is accountable for handling the IoT gadgets along with their operation, as well as operating the security.
  iii. *Services layer* focuses on providing the various necessary services to the consumers, focusing on activities provided on the development side.
  iv. *Communication layer* is used to provide an interface for communicating the information from the sensing layer to the services layer.
   v. *Network layer* assists the gadgets by processing and then sending the information through the internet.
  vi. *Hardware layer* is used for assimilation of various hardware objects required for deployment on the IoT platform.
 vii. *Environment layer* determines the possibility of detecting physical objects (such as moving vehicles, humans, etc.) or other aspects of the environment, e.g., humidity or temperature, of a place etc.

6.3.1.6   Fog and Cloud-based Architecture Layers

In [8], the authors presented a Fog architecture based on [23], which is composed of six layers as shown in Fig. 6.2. In Fog architecture, four layers, namely observing, pre-processing, storing, and security layers, were inserted between the physical layer and the transport layer, before the information is passed to the cloud. As mentioned in [21], the functionalities of these layers can be described as:

    i. *Physical layer* is focused on the examination of objects.
   ii. *Monitoring layer* is responsible for observing the factors, like services and assets, by the consumers, and varied reactions.
  iii. *Pre-processing layer* is responsible for various operations like managing, filtering, and analyzing the information.
  iv. *Storage layer* is accountable for storage and distribution of the gathered information, whenever required.
   v. *Security layer* works on the fundamental security aspects of the information, by preventing it from unauthorized access.

vi. *Transport layer* is the uppermost layer of the fog architecture that allows transmission of data.

Cloud-based architecture is the one which provides the feature of centralized access over the information, using a cloud-based information-handling system, which allows the IoT system to possess a cloud-centric architecture. It is gaining popularity day-by-day in IoT systems because of its intelligent type of sensing of the information and of producing results in the form of information from IoT gadgets. In cloud-based architecture, the cloud is situated between the application and the network layers [21].

6.3.2 DOMAIN-SPECIFIC ARCHITECTURES

6.3.2.1 CloudThings Architecture

In [24], an IoT-based scenario is described which uses the smart home. In this concern, a new architecture, named CloudThings, was presented as a cloud-centric IoT platform. The CloudThings architecture makes use of the following services:

- PaaS: Platform as a Service
- IaaS: Infrastructure as a Service
- SaaS: Software as a Service

Collaboration between the IoT and the cloud denotes a realistic way through which one can easily smoothen the development of an application. Yet, there is a need for dealing with the heterogeneous nature of the IoT physical (or dynamic) things [14].

6.3.2.2 Industrial IoT Architecture

The associated architecture is depicted in Fig. 6.4, which is meant for green Industrial IoT. Its components are:

i. *Sense entities domain*: To increase the energy savings, nodes and smart devices are organized in the sense entities field and they are categorized as:
   - Control nodes (CNs).
   - Sense nodes (SNs).
   - Gateway nodes (GNs).
ii. *Constrained RESTful Network*: RESTful web services are hosted by the network, with the sense entities becoming connected with the Cloud Server.
iii. *Cloud Server*: This provides the Virtual Environment for the objects which are further transmitted to the applications server. It also processes and computes the extracted information.
iv. *User applications*: Through the application server interfaces, clients can interact with the application server, even though they do not have access to the server side codes. The administration node can grant direct access to the clients [25].

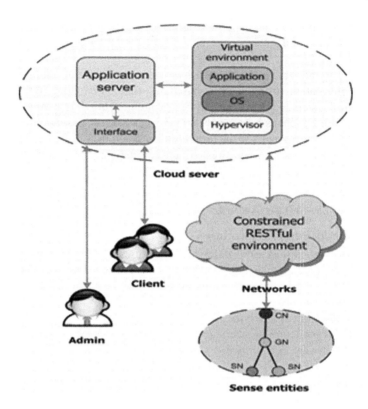

FIGURE 6.4   Energy-efficient IoT architecture [26].

6.3.2.3   Smart and Connected Communities (SCC) IoT Architecture

In [27], the authors proposed an IoT-based SCC (Smart and Connected Communities) architecture. This IoT architecture consists of four different layers, and is specifically designed for smart cities as illustrated in Fig. 6.5.

  i. *Responding layer (or sensing layer)* is the outermost layer of SCC IoT architecture. It is the perceiving layer, with which smart gadgets communicate with the IoT system. It performs the functions of perceiving and transmitting the information further to the interconnecting layers.
 ii. *Interconnecting layer* performs functions similar to those the human nervous system carry out in the human body. It is the connecting layer and its main purpose is data exchange and transfer between different gadgets and diverse domains.
iii. *Data layer* is considered to be the brain of the smart cities, responding to requests from services of the service layer, such as smart lighting system in bad weather, With the assistance of the knowledge base and real-time analysis of data, the intelligent decision system automatically takes the required decisions and carries out the actions. It is used to back-up large amounts

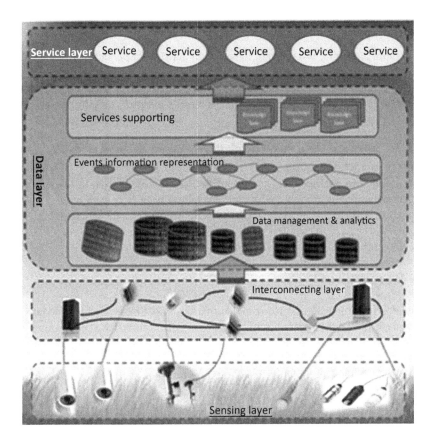

FIGURE 6.5 SCC IoT architecture [27].

  of mixed and insignificant information, and representing it in a significant manner, for knowledge management, decision-making processes, etc..

 iv. *Service layer*, also known as the application layer, provides a variety of essential services for small cities. This layer offers the interface between the IoT system and consumers. The services depend on the service supporting system.

6.3.2.4 3G-PLC (Power Line Communication) Architecture

In [28], 3G-PLC-based IoT architecture is proposed, which collaborates with complex communication systems, namely:

 i. 3G network
 ii. Power Line Communication (PLC) network.

Increasing the number of nodes is the primary objective for using the 3G-PLS network. To accomplish this, IoT framework layers are integrated with the purpose of discovering a new 3G-PLC IoT-based architecture.

Benefits: The advantages of this proposed architecture are:

- Improved services as related to backhaul network opponents.
- Lower cost of network structure.

Limitation: However, this architecture has a major limitation, lacking the incorporation of network heterogeneity [14].

6.3.2.5   Software-Defined Network (SDN)-Based Architectures

In [29], the author designed a SDN-based IoT architecture. The motive behind implementation of this architecture was to enable various IoT tasks in a heterogeneous wireless network environment with high-level Quality-of-Service (QoS).

Benefits: This proposed architecture presents various benefits such as:

- Flexibility.
- Effectiveness in terms of data flow.
- Efficient resource management.

Many existing studies have also revealed that wireless SDN-based architecture can assist in achieving the objectives of IoT in terms of improved scalability, quality of service, context-aware semantic information retrieval, and simple, fast placement of resources.

Limitation: This architecture lacks the organization of the layer created by the controller as it is hard to supervise heterogeneous IoT networks.

6.3.2.6   MobilityFirst Architecture

In [30], a name-specific Future Internet Architecture (FIA), known as MobilityFirst, was introduced by the authors to assist in resolving many problems concerned with cellular (mobile) phones, when these phones were acting as gateways to the Wireless Sensor Networks (WSN) in IoT. The ability of the network is measured and matched to the sensor information speed at a given location.

Benefits:

- *Ad hoc* services
- High security

Limitation: The absence of motivation processes for mobile providers to use the system is a prime limitation [14].

6.3.3   INDUSTRY-DEFINED ARCHITECTURES

6.3.3.1   Cisco Architecture

It follows an Edge model. Fig. 6.6 illustrates the IoT reference architecture and its different levels. The data flow follows a bidirectional approach in IoT:

Internet of things reference model

FIGURE 6.6   Cisco reference IoT architecture [33].

- In a *control pattern*, control information flows from the top to the bottom.
- In a *monitoring pattern*, the control information flows from the bottom to the top.

The processing layer in a 5-layer model is composed of three layers, where data fusion, like gathering information from all the devices, is handled by the data abstraction layer, storage is performed by the data communication layer, and real processing is achieved by the edge computing layer. Level 3 activities aim to carry out large-scale information analysis and transformation. For instance, a sensor device generates the data samples throughout the year at Level 1 a number of times throughout the day. Despite generating this large amount of information, the collected information was never analyzed. Just because of the slow processing of the majority of IoT solutions, analyzing the data will be difficult. Architecture which uses the cloud to carry out analysis will face different problems in the future, because of excessive data.

The approach followed by Cisco is called edge computing. It will solve the issue, by calculating in real time [8, 31, 32], such as, by transmitting the videos from security cameras (as they are worthless) in place of processed video of an event, it will be productive. Another example is in the banking systems, where edge computing is used for detection of fraudulent transactions in ATMs. Cisco promotes its gateways as edge computing devices and it is making trends. Early calculations were also possible with Cisco edge computing; after processing, these analytics are sent to the cloud. Acting on this information, which is near-source, some calculations allow continuous operation, even if the network is not active [8].

6.3.3.2 IBM Architecture

This cloud-computing architecture has been transformed to another model, along with some enhancements, such as device control and management [34]. In that aspect, the layers, in which they are expert with respect to the seven-layer model, is explained here. Much of the IBM architecture for IoT deals with middleware, rather than the whole architecture. Considering the power of IBM's Watsons IoT [35] with

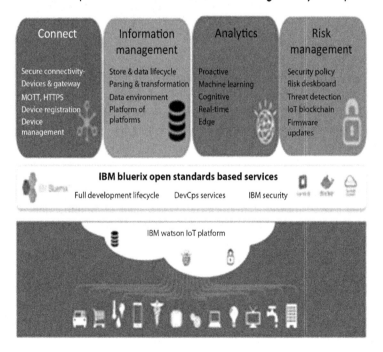

FIGURE 6.7   IBM reference IoT architecture [36].

Cisco's edge computing, as shown in Fig. 6.7, the architecture components are categorized into the following major components:

- *Connect*: The Connect component ensures secure device connectivity (via Gateway), control of gadgets, and communication with third-party services. It is also responsible for device management, as gadgets with device organization agents can execute device organization actions, which can be triggered by the Watson IoT Platform dashboard or the device organization Application Programming Interface (API)
- *Information management*: Information management handles the storing, archiving, metadata management, streaming data, and parsing and managing of the structured and unstructured data. By using the Watson IoT Platform Last Event Cache API, the last event that was sent by a device can be retrieved.
- *Analytics*: This covers a lot of functionality, such as covering social data analytics, machine data analytics, and cultured text analytics for incident data. It is basically intended for applications with millions of records per second, i.e., high-speed applications. Organizations can anticipate more efficiently when there is a greater probability of an event happening by analyzing real-time information from the sensors. In addition, with this high-end data analytics, predictive, cognitive, and real-time data analytics can be carried out.
- *Risk management*: This module ensures leverage of the correct and secure information from the right sources. It performs the functions like e-information protection, auditing, key management, firmware updates, and managing specific security etc. In other words, this layer manages the integrity of the device [8]. By referring to Watson IoT Platform cloud analytics, directive conditions are specified that are dependent on real-time gadget data and that trigger alarms and elective actions when they occur [8, 35].

6.3.3.3 INTEL Architecture

The INTEL System Architecture Specification (SAS) features two versions [37]:

- *Version 1.0*: For Connecting the Unconnected. It allows the system builders to securely connect and manage legacy gadgets, using an IoT gateway,
- *Version 2.0*: For Smart and Connected Things. It specifies how to integrate a variety of smart and connected things, ranging from battery-powered devices to ultra-high-performance devices. It is future- concerned architecture, i.e., it assists seamless cyber physical models.

Fig. 6.8 illustrated a SAS-based layered architecture where:

- The used entity is represented by the white block,
- The run time layer is represented by dark blue blocks and
- The developers are represented by light blue blocks.

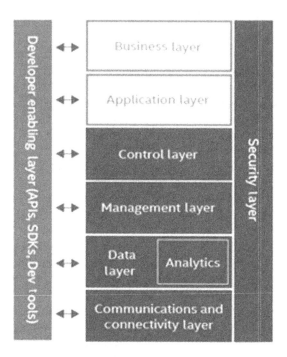

FIGURE 6.8   INTEL architecture [37].

The main objectives of the eight layers shown in Fig. 6.8 are:

- *Control layer*, with the help of policy and control APIs, offers the capability to detach the organization and control. It can take the gadgets away off-premises for cloud-based platform/remote control, which is a basic need of an SDN.
- *Business layer* accesses other layers in the solution through the application layer.
- *Security layer* provides protection to all the layers [8, 37].

6.3.3.4   Microsoft Architecture

The architecture recommended by Microsoft in [38] for IoT applications is cloud native, micro-service, and serverless based. Microsoft recommended that sub-systems communicate over HTTPS/REST, using JSON (as it is human readable), though binary protocols should be used for high-performance needs. The architecture also supports a hybrid cloud and an edge-compute strategy, i.e. some data processing is expected to happen on-premises. It is recommended to use an orchestrator, such as Azure Managed Kubernetes or Service Fabric, to scale individual subsystems horizontally, or PaaS services, e.g., Azure App Services, that offer built-in horizontal scale capabilities.

6.3.3.4.1   Subsystems of Microsoft Architecture for IoT
An IoT application consists of the following subsystems:

- *Devices* (and/or on-premises edge gateways) that are able to securely index with the cloud and offer connectivity options for transmitting and reception of information with the cloud.
- *Cloud Gateway (IoT hub)* is a service to safely receive the information and to offer gadget management options.
- *Stream Processors* that consume the data, assimilate with *business processes*, and save the data into *storage*.
- *User Interface* to envision telemetry data and enable gadget management.
- *Data Transformation* permits reforming, retransforming, and recombination of telemetry data sent from devices.
- *Machine Learning* offers various algorithms, which are executed, based on a dataset of previous telemetry data, so that predictive maintenance scenarios are enabled.
- *User Management* involves part of management which allows different roles and responsibilities among various users by dividing different functionality among them (Fig. 6.9).

6.3.3.5 Google Architecture

Google reference architecture for IoT, introduced by the Google Cloud platform, is shown in Fig. 6.10.

6.3.3.5.1 Basic Components of Google Architecture for IoT

The system is divided into the following three components:

- *Device/Gadget* includes firmware that can directly interact with the world and create a link to a network for interaction among themselves.
- *Gateway* assists different gadgets to communicate with cloud services.

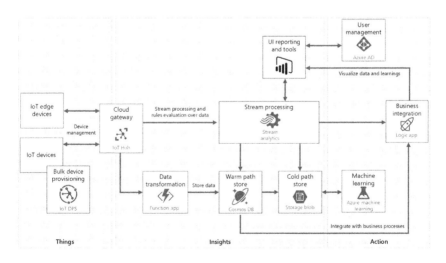

FIGURE 6.9 Microsoft Azure IoT reference architecture [38].

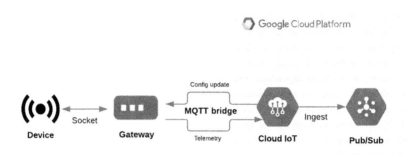

FIGURE 6.10   Google IoT A=architecture [40].

- *Cloud* allows data from a device to be uploaded onto the cloud, where data from different devices, as well as other business-transactional data, are also available for combining and further processing [39].

6.3.3.6   AWS IoT Architecture

Amazon Web Services (AWS) reference Architecture for IoT was a controlled cloud platform in Amazon IoT solutions.

Its main features are:

- It helps web-connected gadgets to communicate with cloud applications and other gadgets securely, thus providing a secure device gateway, authentication, and authorization for proven exchanged data.
- Registration for the identification of gadgets.
- Thing Shadows for storage and retrieval of existing state data.
- Rules engine for communication between gadgets and AWS services.
- Amazon has the most comprehensive service among the cloud providers, so that using the AWS IoT inherently provides you with other benefits.
- With the built-in Kibana integration, this platform becomes highly scalable, resulting in IoT applications that gather, handle, analyze, and respond to information generated. In addition, the on-the-spot registration automatically registers new certificates as part of the initial interaction between the device and AWS IoT, speeding up batch deployments.
- AWS IoT enables integration with the rest of the AWS services, like AWS Lambda, Amazon Dynamo DB, Amazon Kinesis, Amazon S3, and Amazon SNS.
- The AWS IoT offers four different interfaces for creating and interacting with things (Fig. 6.11). These are:
  - i.   Command line interface
  - ii. API
  - iii. AWS SDKs
  - iv. Device SDK [41].

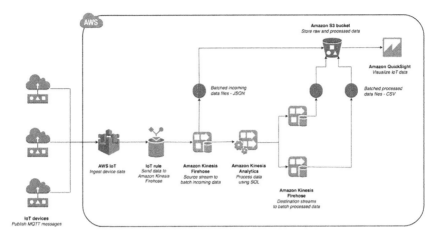

FIGURE 6.11   The AWS IoT platform [41].

6.3.4   Association of Industry-Defined Architectures

The author of [42] carried out a survey to find the top IoT companies. The author collected the data from different sources like company websites, Twitter, IoT Analytics, Google, and LinkedIn, as described in Table 6.2.

Based on the facts and figures represented in the Table, it is quite obvious that Intel had taken over from Google in 2015 in terms of the number of searches. A survey of the number of employees was conducted of companies that carry "IoT" as their tag, so the growth of the top 5 IoT companies was shown in the form of a bar graph (Fig. 6.12).

6.4   CONCLUSION

Much technological advance has been carried out in the field of IoT but still we are lagging behind with respect to a standard architecture. There is no standard architecture defined up to now so, to gain the appropriate facts about IoT architectures, this review presents the state-of-the-art of various architectures, classified on the

TABLE 6.2
Comparison of Top 5 IoT Companies [42]

Company	Searches on Google	Tweets to Twitter	Newspapers and Blogs	LinkedIn survey
Intel	1K	2.6K	4K	616
Microsoft	480	1.6K	26K	545
Cisco	1K	1.4K	5K	719
Google	390	3.1K	21K	99
IBM	720	1.5K	7K	504

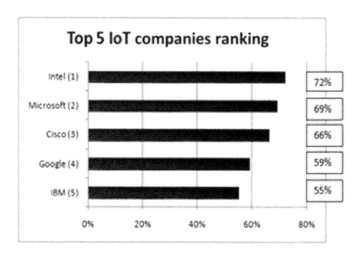

FIGURE 6.12   Top 5 IoT company rankings [42].

basis of the number of layers, domains, and those of industry-defined structures. This paper also compares the industry-defined architectures. However, we are still lacking information on various domains like noise detection, agriculture, etc., which can be explored further.

BIBLIOGRAPHY

1. T.D. Nguyen, J.Y. Khan, and D.T. Ngo. 2016. "Energy Harvested Roadside IEEE 802.15.4 Wireless Sensor Networks for IoT Applications." *Ad Hoc Networks*. doi:10.1016/j.adhoc.2016.12.003.
2. M. Bilal. 2017. "A Review of Internet of Things Architecture, Technologies and Analysis Smartphone-Based Attacks Against 3D printers." *arXiv*, arXiv:1708.04560 [Online]. Available: https://arxiv.org/abs/1708.04560. [Accessed September 22, 2018].
3. International Telecommunications Union. 2005. ITU Internet Reports 2005: The Internet of Things, Executive Summary, Geneva: ITU.
4. I. Ungurean, N. Gaitan, and V.G. Gaitan. 2014. "An IoT Architecture for Things from Industrial Environment." In: *10th International Conference on Communications (COMM)*, Bucharest, pp. 1–4.
5. "Cisco Visual Networking Index: Forecast and Methodology, 2014–2019." Cisco, 2015 May 27. http://www.cisco.com/c/en/us/solutions/collateral/service-provider/ip-ngn-i p-next-generation-network/white_paper_c11-481360.pdf.
6. K. Rose, S. Eldridge, and L. Chapin, 2015 October. "The Internet OF THINGS: An OVERVIEW- Understanding the Issues and Challenges of a More Connected World." *The Internet Society (ISOC)*.
7. Melanie Swan. 2012. "Sensor Mania! The Internet of Things, Wearable Computing, Objective Metrics, and the Quantified Self 2.0." *Journal of Sensor and Actuator Networks*, doi:10.3390/jsan1030217.
8. N.M. Masoodhu Banu and C. Sujatha. 2017. "IoT Architecture a Comparative Study." In: *International Journal of Pure and Applied Mathematics*, 117(8), pp. 45–49. doi:10.12732/ijpam.v117i8.10.

9. N.M. Kumar and P.K. Mallick. 2018. "The Internet of Things: Insights into the Building Blocks, Component Interactions, and Architecture Layers." In: *International Conference on Computational Intelligence and Data Science (ICCIDS)*.

10. J. Lin, W. Yu, N. Zhang, X. Yang, H. Zhang, and W. Zhao. 2017 October. "A Survey on Internet of Things: Architecture, Enabling Technologies, Security and Privacy, and Applications." *IEEE Internet of Things Journal*, 4(5), pp. 1125–1142.

11. H. Suo, J. Wan, C. Zou, and J. Liu. 2012. "Security in the Internet of Things: A Review." In: *Computer Science and Electronics Engineering (ICCSEE), 2012 International Conference on*, pp. 648–651.

12. M. Leo, F. Battisti, M. Carli, and A. Neri. 2014 November. "A Federated Architecture Approach for Internet of Things Security." In: *Proceedings Of 2014 EURO Med Telco Conference (EMTC)*.

13. A. Al-Fuqaha, M. Guizani, M. Mohammadi, M. Aledhari, and M. Ayyash. 2015. "Internet of Things: A Survey on Enabling Technologies, Protocols, and Applications." *IEEE Communications Surveys and Tutorials*, 17(4), pp. 2347–2376. Fourth quarter.

14. I. Yaqoob et al., 2017 June. "Internet of Things Architecture: Recent Advances, Taxonomy, Requirements, and Open Challenges." *IEEE Wireless Communications*, 24(3), pp. 10–16.

15. L.D. Xu, W. He, and S. Li, 2014 November. "Internet of Things in Industries: A Survey." *IEEE Transactions on Industrial Informatics*, 10(4), pp. 2233–2243.

16. P.P. Ray. 2016. "A Survey on Internet of Things Architectures." *Journal of King Saud University – Computer and Information Sciences*. doi:10.1016/j.jksuci.2016.10.003.

17. "Digital Service Innovation and Smart Technologies: Developing Digital Strategies Based on Industry 4.0 and Product Service Systems for the Renewal Energy Sector: Scientific Figure on ResearchGate." Available: https://www.researchgate.net/Service-oriented-architecture-f or-IoT-adapted-from-Li-et-al-2015_fig5_309807447. [Accessed September 23, 2018].

18. R. Khan, S.U. Khan, R. Zaheer, and S. Khan. 2012. "Future Internet: The Internet of Things Architecture, Possible Applications and Key Challenges." In: *10th International Conference on Frontiers of Information Technology*.

19. T. Ara, P.G. Shah, and M. Prabhakar. 2016 December. " Internet of Things Architecture and Applications: A Survey." *Indian Journal of Science and Technology*, 9(45), pp. 0974–5645. doi:10.17485/ijst/2016/v9i45/106507.

20. I.B. Thingom. 2015. "Internet of Things: Design of a New Layered Architecture and Study of Some Existing Issues." *IOSR Journal of Computer Engineering (IOSR-JCE)*, 1, pp. 26–30.

21. N.M. Kumar and P.K. Mallick. 2018. "The Internet of Things: Insights into the Building Blocks, Component Interactions, and Architecture Layers." In: *International Conference on Computational Intelligence and Data Science (ICCIDS)*.

22. D. Darwish. 2015. "Improved Layered Architecture for Internet of Things." *International Journal of Computing Academic Research (IJCAR)*, 4, pp. 214–223.

23. P. Sethi and S.R. Sarangi, "Internet of Things: Architectures, Protocols, and Applications." *Journal of Electrical and Computer Engineering*. doi:10.1155/2017/9324035.

24. J. Zhou et al., 2013. "Cloudthings: A Common Architecture for Integrating the Internet of Things with Cloud Computing." In: *IEEE 17th International Conference Computer Supported Cooperative Work in Design*, pp. 651–657.

25. K. Wang, Y. Wang, Y. Sun, S. Guo, and J. Wu. 2016. "Green Industrial Internet of Things Architecture: An Energy-Efficient Perspective." *IEEE Communications Magazine*, 54(12), pp. 48–54. doi:10.1109/MCOM.2016.1600399CM.

26. Green Industrial Internet of Things Architecture: An Energy-Efficient Perspective: Scientific Figure on ResearchGate." Available: https://www.researchgate.net/figure/ Energy-efficient-IIoT-architecture_fig1_311750469. [Accessed September 23, 2018].

27. Y. Sun, H. Song, A.J. Jara, and R. Bie. 2016. "Internet of Things and Big Data Analytics for Smart and Connected Communities." *IEEE Access*, 4, pp. 766–773. doi:10.1109/ACCESS.2016.2529723.
28. H.-C. Hsieh and C.-H. Lai. 2011. "Internet of Things Architecture Based on Integrated PLC and 3G Communication Networks." In: *IEEE 17th International Conference Parallel and Distributed Systems*, pp. 853–56.
29. Z. Qin et al., 2014. "A Software Defined Networking Architecture for the Internet-of-Things." In: *IEEE Network Operations and Management Symposium*, pp. 1–9, May 2014.
30. J. Li et al., 2013. "A Mobile Phone Based WSN Infrastructure for Iot Over Future Internet Architecture." In: *Green Computing and Communications IEEE and Internet of Things, and IEEE International Conference Cyber, Physical and Social Computing*, pp. 426–33.
31. "The Internet of Things Reference Model, Whitepaper from Cisco." 2014.
32. Z. Javeed. 2018, May. "Edge Analytics, the Pros and Cons of Immediate, Local Insight." [Online]. Available: https://www.talend.com/resources/edge-analytics-pros-cons-im mediate-local-insight/. [Accessed September 22, 2018].
33. "Development of an Architecture for a Tele-Medicine-Based Longterm Monitoring System: Scientific Figure on ResearchGate." Available: https://www.researchgate.net/The-visual-representation-of-Internet-of-Things-Model-shows-7-layers-from-1-the_f ig8_315892029. [Accessed September 23, 2018].
34. A. Gerber. 2017 August 7. "Simplify the Development of Your IoT Solutions with IoT Architectures, IBM Document."
35. "Watson IoT Platform Feature Overview, IBM Cloud." 2018 [Online]. Available: https://console.bluemix.net/docs/services/IoT/feature_overview.html#feature_overview.
36. "Webcast: IoT Use-Cases with IBM Watson IoT Platform." *Mario Noioso*, 2016 May 11. Available: https://marionoioso.com/2016/11/05/webcast-iot-use-cases-with-ibm-watson-iot-platform/.
37. "The Intel IoT Platform, Architecture Specification White Paper."
38. "Microsoft, Azure IoT Reference Architecture, Microsoft." 2018. Available: https://azure.microsoft.com [Online]. https://azure.microsoft.com/en-us/blog/getting-started-with-the-new-azure-iot-suite-remote-monitoring-preconfigured-solution/.
39. "Overview of Internet of Things." 2018 Available: https://cloud.google.com/ [Online]. https://cloud.google.com/solutions/iot-overview [Accessed September 29, 2018].
40. "Google Cloud IoT." 2018 Available: https://cloud.google.com/ [Online]. Available: https://cloud.google.com/solutions/iot/ [Accessed September 29, 2018].
41. "9 Platforms That Have Revolutionised the World of Things Connectivity—Part 1." [Online]. Available: https://itnext.io/9-platforms-that-have-revolutionised-the-world-of-things-connectivity-part-1-2664624d6f65. [Accessed September 29, 2018].
42. K.L. Lueth. 2015. "The top 20 Internet of Things companies right now: Intel overtakes Google." *IoT Analytics*, February 24. [Online]. Available: https://iot-analytics.com/20-i nternet-of-things-companies/.
43. Kavita Taneja, Harmunish Taneja, and Rohit Kumar. 2018. "Multi-Channel Medium Access Control Protocols: Review and Comparison." *Journal of Information and Optimization Sciences*, 39(1), pp. 239–247.
44. Harmanjot Kaur and Rohit Kumar. 2019. "A Survey on Internet of Things (IoT):Layer-Specific, Domain-Specific and Industry-Defined Architectures." *Advances in Computational Intelligence and Communication Technology*, pp. 260–270.

7 Hybrid Energy Systems

Akhil Gupta, Kamal Kant Sharma, and Akhil Nigam

CONTENTS

7.1 INTRODUCTION

In today's times, renewable energy systems (RESs) can replace the use of fossil fuel with the installation of the latest renewable sources. There are many RESs available, each type performing best under particular climatic conditions, but, due to the unpredictability of the availability of any of the energy sources, the performance of the overall system may be disturbed [2–4]. A single RES may not be able to fulfill all the demands of the end-user,

A hybrid energy system is a combination of two or more systems, connected together in order to provide maximum deliverables in terms of energy and power, to provide sufficient cooling, heating, and hot water for domestic buildings and industrial applications [1], as depicted in Figure 7.1. By employing a hybrid RES, combining two (or more) RESs, if one energy source is unavailable or unable to perform as per requirements under certain climatic conditions, the second source fulfills the requirement and provides an appropriate response [2–4]. A combination of RESs, connected together, would provide a better and more reliable output than a single RES, so these hybrid systems have been preferred for many years [3]. Various sources have been used, with different configurations and approaches, to develop new arrangements to achieve faster energy supply, with better deliverables in terms of efficiency. With a hybrid RES, various sources, which are similar in their operation or which have similar properties, can share the same cluster requirements and overcome the necessity for additional manpower and for sophisticated equipment for energy storage.

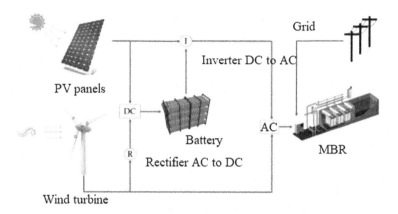

FIGURE 7.1   Schematic diagram of a typical hybrid energy system

A hybrid RES can be seen as an alternative arrangement to conventional energy sources, resulting in a reduction in harmful gas emissions and an increase in sustainability indices [5, 6]. Various case studies have proved that, in hybrid systems, multivariables are not preferred over different set of variables. For example, in the case of RES, combinations of only non-conventional energy sources will increase the order of a system but the degree will remain the same; as a consequence, a hybrid system will increase the order and degree of a system in such a way that balanced algorithms can be expressed together, to control the output in a range. Hybrid RESs, in which RESs of different kinds are combined together and split into different arrays, also alter the myth that RESs are not cost effective; most hybrid RESs can combine conventional energy sources together with non-conventional energy sources, resulting in microgrid and whole-energy systems, which will work as base power plants and peak power plants simultaneously. These kinds of terminologies provide enough space for emerging technologies to grow, be cost effective for investors, and be easy to implement at remote locations. Hybrid RESs may be a combination of conventional and non-conventional energy sources [3, 7–9].

Hybrid RESs also operate in different configurations, namely traditional or emerging mode. In traditional mode, all the sources connected together will operate together simultaneously and the grid operating in such a manner will operate as a base power plant but will not be economical as it will not be more effective than conventional energy sources. For example, if solar and wind systems are combined together and work simultaneously, the two systems will work together but their inputs will be variable and intermittent in nature and hence not reliable. On the other hand, if one member of a system fails, then the other system is also liable to fail, resulting in collapse of the entire system.

The other mode of operation is emerging mode, which is also known as sequential mode. In sequential mode, the sources connected together do not operate together ("simultaneously") but in a sequential manner. This type of operation is better than the traditional mode configuration but limited to operate only as peak power plants, being cost effective with a load ("demand") factor <100%. Due to these disadvantages, hybrid energy systems need to operate in a sequential mode, when the two systems are of different nature, such as conventional and non-conventional energy sources, resulting in one of the systems operating as a base power plant, whereas the other will operate as a peak power plant. In hybrid power systems, one source will be added to the system, to act as a backup for one member of the system, to make the hybrid system effective. Hybrid renewable energy (RE) systems are composed of at least one conventional and another, non-conventional energy source [10–12]. Due to drawbacks associated with current hybrid RE systems, many research groups are forming and simulating integrated circuits, consisting of non-conventional energy sources, in combination with tested sources, in connected traditional grid codes and a deregulated environment, to provide customers with competition among power suppliers. In India, a deregulated environment is limited to only individual users or a single organization, whereas, in other countries, especially in Europe, deregulated environments are evaluated and analyzed with multiple attributes for different end-users. These possibilities lead to better networks and greater reliability [13].

In remote and isolated areas, the uses of hybrid energy systems are more widely preferable than those of grid-connected systems [14, 15]. Along with distributed generation or microgrid systems, the use of hybrid systems has been popular for a number of years and has become a topic of great interest [3]. Various studies have analyzed hybrid energy systems on the basis of different operating characteristics and by evaluating significant parameters within the hybrid operation. Certain parameters, which seem to be important in RES, are not relevant to conventional energy sources, although many parameters are common to both RES and conventional systems. These are studied with microgrids following smart grids requirements with requisites for hybrid systems [7]. The advances in power electronic converter technology and in automatic controllers have resulted in improvements in the systems performance of the hybrid system, making them more reliable, efficient, and economical [16–18].

7.2 DESIGNING A HYBRID ENERGY SYSTEM

Hybrid systems are complex energy systems, with many sets of variables; they have shared different resources of electric power and are connected by a controller for parallel distribution to load feeders. The main concern is suitability and reliability in terms of the processing unit, the processed and unprocessed parameters, and the size of the system. The size of the hybrid system determines the number of computations and iterations required in a stipulated processing time to evaluate the optimum result. A hybrid system also involves feedback (closed and open) systems required to obtain the desired output, with real and reactive power controls over a limited range, so that frequency mismatching and voltage stability issues can be assessed well ahead of time. Moreover, when the number of resources connected together, with different operating characteristics (such as hydro and wind), then the various issues for power quality of the individual resource are evaluated. In hybrid systems, the number of devices which operate the real and reactive power controls and provide the desired output, with controlled deliverables, are commonly termed "power conditioning devices", involving the use of flexible alternating current transmission system (FACTS) devices with power electronic interfaces.

Different combinations of energy sources, which can be developed into a system with suitable sustainability indices, also depend upon the availability of sufficient land and access to resources necessary to make the planned hybrid system [19]. RES has also gained importance in hybrid systems due to its many advantages. A major agreement, signed between developing and developed countries and termed the Kyoto Protocol, signifies that the emission of greenhouse gases and other harmful gases should be avoided or reduced, with the benefit of carbon credits, evaluated in dollars per unit, being paid to developing countries by developed ones. These kinds of agreements and interactive knowledge-sharing between these two country types across the globe allow researchers to develop simulation models of hybrid systems, exhibiting environmental and economic benefits. These parameters help to achieve proper modeling designs of hybrid RE systems [1].

Some guidelines need to be followed for the design of hybrid energy systems:

- Type of RE to be installed.
- Number and estimated capacity of RE units.
- Energy storage system, if necessary.
- Decision as to whether the system is standalone or traditional (grid code) mode

Before developing simulation models and incorporating advances in new technology into the traditional system, the planning needs of the hybrid system, with respect to software implementation and availability of resources, are analyzed to avoid bottlenecks and to implement ATC (Available Transfer Capability) when considering RES for a selected location. RES also depends upon climate changes, as these parameters are natural resources. Therefore, climatic conditions play a very important role in making particular decisions. Potential benefits of a hybrid energy system include greater reliability, prolonged, sustainable, and low-cost services, freedom from pollution, efficiency and ease of installation, and the generation of clean and green energy.

7.3 CLASSIFICATION OF HYBRID ENERGY SYSTEMS

There are many possible configurations of hybrid energy systems, which operate differently under the influence of numerous climatic conditions [20, 21]. These hybrid energy systems are classified as:

- Hybrid wind-solar system
- Hybrid diesel-wind system
- Hybrid wind-hydropower system
- Hybrid fuel cell-solar system
- Hybrid solar-thermal system

7.3.1 HYBRID WIND-SOLAR SYSTEM

Hybrid systems are complex and many issues are associated with them but they have been developed considering the availability of and need for a particular type of land. Researchers simulating hybrid systems involve renewable energy sources, such as wind and solar, due to their abundance in nature and their interdependence, but both of these systems are intermittent in nature, and are unable to fulfill the load duration curve and to cater for the load uniformly. As a result, neither energy source can operate individually as neither alone will be able to support the capital costs incurred, whereas integrated systems lead to power quality issues as well as voltage sag and swell for a prolonged duration of time. Combined wind and solar systems can be used only at coastal regions or for a small population where the load is fixed and can be predicted in advance [17]. A schematic of a typical hybrid solar-wind system is depicted in Figure 7.2.

There are various issues pertaining to hybrid solar-wind systems which need to be considered, namely solar irradiance, solar orientation, annual average wind speed, total output predicted, average temperature, and land suitability [18]. Both systems need to cater for the load from end users and must be tested in both islanding mode and grid-connected mode, so that, should any element of the system ever fail, the secondary systems can fulfill the minimum requirements (base load), and the grid-connected mode should be available, as both systems are unpredictable. Grid codes must be evaluated for a specific region where hybrid systems are planned in consideration of energy requirements. Hybrid systems must follow grid code requirements, and operation should be in parallel, so that voltage stability is not affected, with extra energy being stored in batteries, to ameliorate economic aspects.

7.3.2 HYBRID DIESEL-WIND SYSTEM

There are two diesel-wind strategies which can be followed: the first one is to operate a diesel engine continuously, and the second one is to operate the diesel engine at selected time intervals on a time scale of 4 out of 5 or on an intermediate basis, as required. Important components required are load, generators with a power

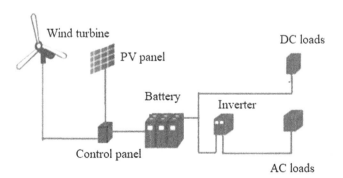

FIGURE 7.2 Schematic diagram of a typical hybrid wind-solar system

electronic controller, a diesel engine, and a battery energy storage system [13]. A schematic of a typical hybrid diesel-wind system is depicted in Figure 7.3. In a diesel-wind hybrid system at the same time scale, both components operate as peak power and the main objective is to use both systems economically, so that deliverables, in terms of output and efficiency, are cost effective, and optimization can be achieved with a minimal number of iterations. Such diesel-wind hybrid systems are used at remote locations and for small populations on an island or in remote places, where electrification is required in terms of security concerns and economic issues.

7.3.3 HYBRID WIND-HYDROPOWER SYSTEM

This kind of system is commonly used for generation of electricity on a larger scale. Such hybrid systems are very rare, as hydropower is limited by certain requirements, such as a water catchment area (dam and reservoir), whereas wind availability at a particular speed is another constraint, which can vary from place to place. Both systems (wind and hydropower) are location specific and are subjected to various constraints which cannot be overcome by including an alternative energy source. The most promising advantage of this kind of hybrid system involves a combination of a base power plant and a peak power plant, with the hydropower plant acting as the base power plant and the wind power plant behaving as the peak power plant, with the two systems able to complement one another. A pumped-storage plant can be implemented, in which both energy systems will operate in peak mode, but land constraints will still exist [19]. A schematic of a typical hybrid wind-hydropower system is depicted in Figure 7.4.

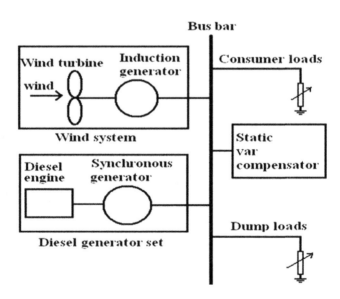

FIGURE 7.3 Schematic diagram of a typical hybrid diesel-wind system

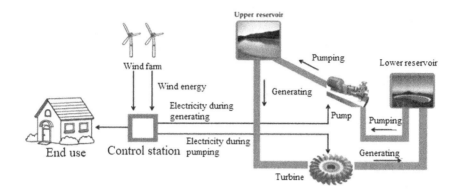

FIGURE 7.4   Schematic diagram of a typical hybrid wind-hydropower system

7.3.4   HYBRID FUEL CELL-SOLAR SYSTEM

Considering a hybrid fuel cell-solar system, input power varies continuously with time, because sunlight irradiance changes continuously. This hybrid system consists of a solar photovoltaic (PV) cell, a proton-exchange membrane fuel cell, and power conditioning devices. The system must be designed in such a way that it performs continuously under ever-changing weather conditions. Electricity is produced by a PV cell to meet consumer demand [20]. When there is any problem regarding generation due to low solar irradiance, the proton-exchange membrane fuel cell is used to maintain the reliability of the system. It is a hybrid system and both systems act as a catalyst to drive the other. One important feature of the fuel cell-solar system is its lower complexity, but it has a drawback of lower reliability indices. It also has a relatively low environmental impact. The power conditioning unit is a bidirectional power converter, which changes DC energy into AC energy; it also becomes a rectifier, changing AC power into DC power. Hence, the power conditioning unit behaves as the heart of the hybrid PV and fuel-cell system. This hybrid system reduces the usage of the fuel cell and improves the voltages of the entire system on a large scale. A schematic of a typical hybrid fuel cell- solar system is depicted in Figure 7.5.

7.3.5   HYBRID SOLAR-THERMAL SYSTEM

Another hybrid system can be operated by combining a solar PV and a thermal system [22]. In this system, the performance of the entire system can be enhanced by variable levels of solar irradiance and the temperature of the plate. This provides both thermal efficiency and electrical efficiency. These efficiencies define the system performance under different climatic conditions [23]. Optical concentration technologies are employed widely with the latest technologies and can be used in solar thermal systems for producing thermal energy storage for industrial application purposes. These kinds of system complement each other but a major drawback is the conversion technology, which is remote and limited, and cannot be accessed very easily by the end user. In addition, from the investor side, the technologies used are

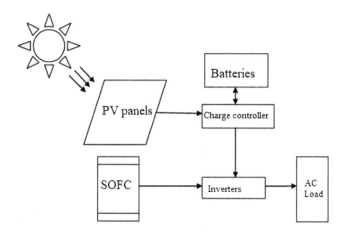

FIGURE 7.5   Schematic diagram of a typical hybrid fuel cell-solar system

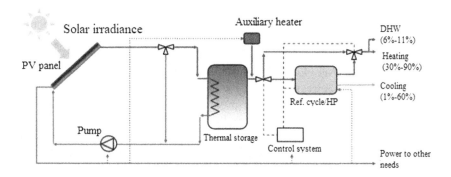

FIGURE 7.6   Schematic diagram of a typical hybrid solar-thermal system

not user friendly. The popular concentrating solar thermal systems, such as parabolic trough collectors, linear Fresnel reflector systems, central receiver systems, and dish systems, are utilized in various applications [21]. A schematic diagram of a hybrid solar-thermal system is depicted in Figure 7.6.

7.4   CONCLUSION

Hybrid energy system installation is good for system performance under variable climatic conditions. It is also beneficial for remote domestic and industrial purposes. Because sometimes only one system is able to perform well, alternative sources of energy are needed to maintain continuity of electric power availability. Moreover, it has also been found that the electric power generation from hybrid energy is easy and safe for the environment. The operating systems have a longer lifespan than other conventional sources of available energy. In this way, hybrid systems can provide more affordable solutions, with greater efficiency for power generation in future.

REFERENCES

1. Md. Ibrahim, A. Khair, and S. Ansari, "A review of hybrid renewable energy systems for electric power generation", *International Journal of Engineering Research and Applications*, vol. 5 no. 8, pp. 42–48, 2015.
2. W. Sparber, K. Vajen, S. Herkel, J. Ruschenburg, A. Theur, R. Fedrizzi, and M. D Antoni, "Overview on solar thermal plus heat pump systems and review of monitoring results", In: *Proceedings of ISES Solar World Congress*, Germany, September 2011, pp. 1–12.
3. P. Ganguly, A. Kalam, and A. Zayegh, "Solar-wind hybrid renewable energy system: Current status of research on configurations, control, and sizing methodologies", *Hybrid-Renewable Energy Systems in Microgrids*, Woodhead Publishing Series in Energy, pp. 219–248, 2018.
4. W. Sparber, and R. Fedrizzi, "8-Hybrid systems for renewable heating and cooling", *Renewable Heating and Cooling Technologies and Applications*, 2016, pp. 181–195.
5. International Energy Agency (IEA), Solar Heat Worldwide, 2015, http://www.iea-shc.org/solar-heat-worldwide.
6. Frequently asked questions about district heating and cooling, Euro Heat & Power Newsletter, May 2016, http://www.euroheat.org/District-Energy-Explained-210.aspx.
7. M. Nehrir, C. Wang, K. Strunz, H. Aki, R. Ramakumar, J. Bing, Z. Miao, and Z. Salameh, "A review of hybrid renewable/alternative energy systems for electric power generation: Configuration, control and applications", *IEEE Transactions on Sustainable Energy*, vol. 2, no. 4, pp. 392–403, 2011.
8. S. Rehman, and K. Tam, "Feasibility study of photovoltaic-fuel cell hybrid energy system", *IEEE Transactions on Energy Conversion*, vol. 3, pp. 50–55, 1998.
9. C. S. Solanki, *Solar Photovoltaics-Fundamentals, Technologies and Applications*, 2nd ed., PHI learning private limited, 2015.
10. Y. J. Reddy, Y. V. P. Kumar, K. P. Raju, and A. K. Ramsesh, "Retrofitted hybrid power system design with renewable energy sources for buildings," *IEEE Transactions on Smart Grid*, vol. 3, no. 4, pp. 2174–2187, December 2012.
11. Y. J. Lee, D. H. Han, B. J. Byen, H. U. Seo, G. H. Choe, W. S. Kwon, and D. J. Kim, "A new hybrid distribution system interconnected with PV array," In: *Proceedings of 7th International Power Electronics and Motion Control Conference (IPEMC)*, Harbin, China, June 2012.
12. Y. J. Reddy, K. P. Raju, and Y. V. P. Kumar, "Real time and high fidelity simulation of hybrid power system dynamics," In: *Proceedings of IEEE International Conference on Recent Advances in Intelligent Computational Systems*, Kerala India, September 2011.
13. A. Testa, S. D. Caro, and T. Scimone, "Optimal structure selection for small-size hybrid renewable energy plants," In: *Proceedings of 2001 14th IEEE European Conference on Power Electronics and Applications*, Birmingham, UK, 2011.
14. G. Bhuvaneswari, and R. Balasubramanian, *Hybrid Wind-Diesel Energy Systems*, Woodhead Publishing Limited, pp. 191–215, 2010.
15. M. M. Balasubramanian, and S. C. Tripathy, "Self-tuning control of wind-diesel power systems", In: *Proceedings of IEEE Proceedings of International Conference on Power Electronics, Drives and Energy Systems for Industrial Growth*, New Delhi, India, vol. 1, pp. 258–264, 1996.
16. G. M. Masters, *Renewable and Efficient Electric Power Systems*, Wiley-Interscience, 2004.
17. R. J. Wai, C. Y. Lin, and Y. R. Chang, "Novel maximum power extraction algorithm for PMSG wind generation system," *IET Electric Power Applications*, vol. 1, no. 2, pp. 275–283, 2007.

18. G. Notton, *Hybrid Wind-Photovoltaic Energy Systems*, Woodhead Publishing Limited, pp. 216–253, 2010.
19. A. S. Ingole, and B. S. Rakhonde, "Hybrid power generation system using wind energy and solar energy", *International Journal of Scientific and Research Publications*, vol. 5, no. 3, pp. 1–4, 2015.
20. O. A. Jaramillo, O. R. Hernández, and A. F. Toledo, *Hybrid Wind-Hydropower Energy Systems*, Woodhead Publishing Limited, pp. 282–322, 2010.
21. A. K. Damral, "Hybrid power generation using PV and fuel-cell", In: *Proceedings of 1st International Conference on Large-Scale Grid Integration of Renewable Energy*, India, pp. 1–7, 2017.
22. M. B. Michael, and E. T. Akinlabi, "A review of solar thermal systems utilization for industrial process heat applications," In: *Proceedings of World Congress on Engineering and Computer Science (WCECS)*, San Francisco, USA, vol. 2, pp. 1–5, October 2016.
23. C. Sharma, A. Sharma, S. Mullick, and T. Kandpal, "A study of the effect of design parameters on the performance of linear solar concentrator based thermal power plants in India," *Renewable Energy*, vol. 87, no. 1, pp. 666–675, 2016.

8 Power Quality Analysis of a Wind Energy Conversion System Using UPFC

Akhil Gupta, Kamal Kant Sharma, Sunny Vig, Gagandeep Kaur, and Ashish Sharma

CONTENTS

8.1 INTRODUCTION

From the previous century, people have started using wind as a major resource for generating electric energy [1]. Altogether, along with hydro-power and photovoltaic power, electric power generated from wind power technology has become one of the most-utilized sources of electric energy. At the end of the 19thcentury, the first laboratory experiment was successfully carried out to use windmills for the generation of electric power. Since then, wind power has become a reliable and cheap source of electric power. This is due to the fact that the several different types of generators can be used for wind electric power generation (WEPG), namely asynchronous generators and induction generators [2–4]. However, stability of electricity supply is the critical issue related to power quality (PQ), due to the fact that wind direction cannot always be maintained constant at all times during the daytime. Therefore, to improve the stability of the power system, flexible AC transmission system (FACTS) devices, also known as power conditioning devices, have seen tremendous exponential growth in implementation in recent years. Basically, WEPG has been treated as a negative load. However, when any type of fault occurs in the system, the wind turbine (WT) is disconnected from the system. PQ issues related to power generation stability in any electrical system are sag-swell, harmonics, power outage, frequency variation, and transients [5, 6].

The above reasons cause instability in the system. A voltage outage occurs when the supply of electric power is interrupted. In power systems, when the voltage of any equipment operates above or below a specified standard range, a fluctuation occurs, which causes the introduction of harmonics and transients into the system. The PQ of electric power from WEPG is increased by using the latest available controllers [7]. Wind and hydro-electric power plants together give efficient results vis-à-vis stand-alone systems. However, WEPG has advantages which are enumerated below.

- **Clean and endless fuel** – WEPG does not produce any emitting pollutant emissions.
- **Financial development at local level** – WT can be a good source of income for a landowner who makes the land available on lease for the development of wind technology.

8.2 MODEL OF A WIND ELECTRIC POWER GENERATOR

In this section, the description of a model for WEPG, capable of harnessing electric power of 5000 MVA and connected to another 1000 MW wind power plant, is presented. This system has been tested for faults in the system, and faults at high frequency have also been simulated. The main objective is to test the system stability in face of the common cases of fault occurrence. Single-phase faults of its property of connecting the line to the ground in case of a live system is termed a(L-G) fault, and have been simulated and the stability has been checked. It has been observed that, with the occurrence of a fault, the FACTS device incorporates and improves the voltage profile in the system as the fault induces current and terminal voltage dips by a few values. To control the voltage profile and to prevent it from dipping by less than 1 p.u., reactive power is injected into a system by unified power flow controller (UPFC). A synchronous generator has been used for wind power generation, because most of the WTs use an induction generator for the generation of AC. Figure 8.1 depicts the operation of asynchronous generator for WEPG. When its rotor runs faster than normal, the induction generator generates current. The slip of the generator is basically the difference between its normal speed and its operating speed [8]. The current is produced in the rotor due to stator flux, when its prime mover rotates the rotor above its synchronous speed. With continuous movement of the rotor and stator magnetic field, the flux has been reduced by a significant value, and coils at the stator side experience current as the coils are connected with the

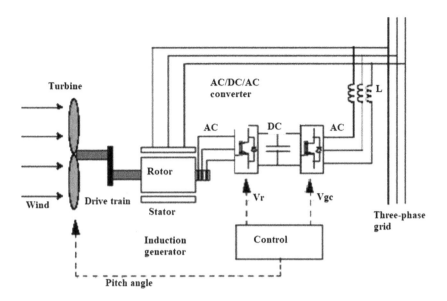

FIGURE 8.1   Schematic representation of an induction generator for wind electric power generation [9]

utility grid. The main underlining objective is to operate the motor as a generator as would happen due to the value of slip becoming negative as the rotor speed exceeds synchronous speed. Similarly, it could be observed in the case of the Ferranti effect occurring in transmission when the lines are lightly loaded.

Analyzing the operation of the induction generator, stator coils are connected with the utility grid, so that current is induced, whereas the rotor is connected to power electronic interfaces to invert the output in a desired form. Connected power electronic interfaces convert AC-DC-AC through converters; the whole operation can be constituted by using choppers but not at a significantly lower voltage value; due to the intermediate processes of conversion, harmonics are developed, resulting in heating and imbalance in the system. Converters give the supply back to the AC utility grid through transmission lines, using coupling transformers. The main advantages of using induction generators are that the operation is very reliable, inexpensive, overloading capacity is good, and the size of the generator is compact.

8.3   MAXIMUM POWER POINT TRACKING (MPPT) FOR WEPG

Renewable energy systems have been simulated because of problems associated with the use of fossil fuels for the generation of electricity. Although renewable energy systems are difficult to integrate, the wind energy conversion system (WECS) has been used extensively due to the abundance and intermittent nature of wind which makes it suitable for peak power plant operation, and also for use in remote locations. Energy production harnessed from wind power changes continuously due to variation in wind direction and wind speed on a continuous basis. Basically, there are three types of MPPT techniques which are given below:

- Power signal feedback control.
- Tip-speed ration control.
- Hill-climb search control.

Tip-speed ratio generally controls the rotating speed of a generator. Using a simulation of power system feedback control, WT curves and power curves at the maximum point can be obtained. Hill-climb search technique is used for finding maximum power points of WT. Unlike the deployment of various search techniques, the main objective of every technique is to find the accurate operating point, which is fully responsible for the value of the power output and the speed of the wind turbine. However, the location and type of the wind turbine depends on the operating point of the wind power plant. The underlying principle lies in a fact that wind flows due to solar irradiance and orientation on uneven surfaces, which make surfaces brighter and hotter. Therefore, wind flows but changes its direction and intensity; to maintain and find the exact intensity and value of the wind, certain algorithms have been developed. Of those algorithms, maximum power point tracking (MPPT) is used to determine the maximum output as a deliverable, which can be obtained from a wind turbine (horizontal or vertical), with particular specifications[10].

This technique involves automated systems with power electronic interfaces, in such a way that output is driven at the maximum value corresponding to its indices, as MPPT involves tracking the power value in terms of its maximum value at output terminals. This kind of technique can also control the output of the wind turbine using its mechanical components, which are fixed but simulated for a particular angle to achieve maximum output in terms of power from a specified wind turbine, which can be custom designed. Moreover, this technique is beneficial in making an automated system but the main drawback is that it creates harmonics, as a result of its operation which has a power electronic interface in the form of converters; these convert AC to DC, and outputs as AC, covering three intermediate processes. Along with this, the technique is costly and requires a sophisticated microcontroller system for a designed operation, which is simulated for a stipulated time period [11, 12]. Various other techniques are implemented, along with MPPT, to control the speed of the wind turbine with pitch angle variation taking into account the magnetic field developed by the doubly fed induction generator (DFIG) or the self-excited induction generator (SEIG), which are different kind of generators used in the generation of electricity in wind power plants. In wind power plants, the main problem is tracking the maximum power and controlling the pitch angle; therefore, automated and intelligent techniques are also used along with heuristic algorithms used in the plant dynamics. Despite this, these techniques are unable to control the ratio of load current to the sum of load current and no-load current to a certain extent; this is due to the fact that, in wind power plants, only power electronic interfaces with buck-or-boost configurations are used.

For a thorough study, simulation of the model can be performed using power-electronics simulation (PSIM) software. Using tip-speed ratio control for a 3 kW wind turbine, the output power follows the C_p curve. The Perturb and Observe (P and O) MPPT technique also compares previous output power. These techniques possess both dynamic characteristics and stable steady-state characteristics, and results obtained show the efficiency of the system by implementing the MPPT technique in achieving improvements of 12%.

Two controlling methods were compared, using the MPPT technique, through which speed control mode adjusts the rotor speed, whereas power control mode adjusts the active power [13]. However, most of the MPPT techniques had been applied on variable speed WTs for maximum wind. In order to regulate the speed of the rotor, the MPPT technique includes both speed and power of the wind turbines in order to match the specifications and derive a relationship between two independent parameters; however, these parameters are recorded and analyzed for PQ improvements. Power control deals with power flow dynamics of the wind turbine and enables a system of the wind power plant to ensure both stability and maximum power output. The power control mode enhances the capacity of the wind turbine to make use of wind speed optimally so that deliverables can be achieved. Basically, operation of wind turbines can be classified on the basis of the converter used, such as a full-scale converter or a partial-scale converter. The main difference in operation involves a change in the duty cycle of the converter, irrespective of wind speed and the independent parameters involved in the plant. Fixed-speed wind turbines employ DFIG or SEIG, but variable-speed turbines make use of mechanical gears to vary the speed so that the maximum output can be obtained. For example, if wind speed is not optimal, at low wind speed, the maximum output should be obtained instead of wasting the minimum speed of the wind. The main function of the controller is to make maximum use of every (uncontrolled)input, like wind speed. In addition to this, the wind power plant also used the MPPT technique for controlling the angular velocity of rotors in the non-utility mode. These types of wind power plants employ variable wind turbines and use synchronous generators, instead of induction generators, because of the fact that the permanent magnet synchronous generator (PMSG) fulfills the purpose of controlling speed, with the fixed wind speed being maintained at the input side, These plants are expensive and complex as they require a converter at the machine side, along with a filter for reducing harmonics, developed using power electronics like DC converters. The controlled converter at the output side, enhancing grid parameters, controls the reactive power of the system. Pulse code modulation (PCM) gives better results for PQ and the system becomes more stable in operation.

The static volt-ampere reactive (VAR) compensator replaces the fixed compensation techniques, using capacitors, implanted variable series, and shunt control of the reactive power. Although it has been observed that these variable compensation techniques also induce stability in a system, RES contains intermittent inputs and generates fixed variables, which are similar to wind power plants [14]. In centralized or decentralized power systems, security and stability are major concerns for wind power plants having multiple turbines, with reactive control and compensation available at load-side of the plant, so that every wind power plant employs FACTS devices in their intermittent and variable operation of WTs [15]. In recent times, wind power has played a crucial role in power systems. Static synchronous compensator (STATCOM), with DFIG, is implemented as a controller which can improve stability in wind voltage. Doubly fed induction generators can be employed with back-to-back converters, which can regulate the flow of active and reactive powers. STATCOM are cascaded at point of common coupling (PCC), through which stability can be enhanced. In this way, the system becomes more flexible and stable [16].

Various case studies has been established in the context of FACTS devices, and results have shown that FACTS device parameters are not controlled and variable, so that various heuristic algorithms and artificial intelligence or trained techniques can be used, like PSO (Particle Swarm Optimization),for multi-machine systems. While implementing PSO at various sides of the test system, care should be taken that the system must have multi-machine characteristics, as PSO is variable dependent and prone to errors for single-machine variable systems, in accordance with wind turbines. Static VAR compensator (SVC) can also be implemented for stability in terms of voltage for a two-machine system, using power system stabilizer (PSS). The underlying principle is to control the magnitude of voltage change (sag or swell) during the fault conditions considering balanced and unbalanced. During the faulty conditions, stability gets affected during switching which involves unwanted voltage and current transients for a short period of time, leading to collapse of the system. These kinds of system are prone to instability, and analysis of the IEEE-9 bus system has been discussed [17, 18].

8.4 FLEXIBLE AC TRANSMISSION SYSTEMS (FACTS) CONTROLLERS

The electrical power system is a complex system and, on the basis of operation, can be classified into one of three different categories: generation, transmission, and distribution. On the distribution side, load is connected and voltage is reduced, due to this reactive power control not being required, whereas real power control is not required as it will be consumed by end-users. Generation is achieved by conventional or non-conventional energy sources by any means, making use of generators which have real and reactive power ratings. But the rating of generators are governed by their loading factor, withstand capacity limits. On the transmission side, voltage is high and classified on the basis of the types of transmission line used, but, whereas voltage is high, reactive power is high and needs to be controlled by absorbing or generating it in a line. Due to these requirements, power electronic interfaces controlled by the triggering angle of silicon controlled thyristor (SCR) (thyristor) have been researched and classified as series or shunt compensators, for compensating reactive power and keeping real power within controlled limits. The real and reactive power of wind power systems are controlled by any type of FACTS controller. Various FACTS controllers are classified according to generation based, like their advances, in a period of time: as an example, first-generation controllers include SVC, second-generation FACTS, unified power flow controller (UPFC), static synchronous series compensator (SSSC), and interline power-flow controllers (IPFC) [19]. Some FACTS controllers are series connected and some are shunt connected. First-generation FACTS controllers are shunt connected and used for parallel compensation at the point of coupling and to maintain voltage profiles in a system by injecting reactive power, whereas second-generation FACTS controllers are series connected and absorb reactive power, so that the voltage can be enhanced among elements of the transmission line. With the advances in FACTS controllers, it has been observed that load flow analysis dynamics improve, while real power and reactive power interactions keep the system stable.

8.4.1 CLASSIFICATION OF FACTS CONTROLLERS

Complete classification of FACTS controllers lists them as UPFC, IPFC, SSSC, static synchronous compensator (SSC), and SVC. These controllers inject power into the AC transmission line. These controllers are classified into two types based on their construction, as either shunt or series [20]. Figure 8.2 shows a summary of FACTS controllers, which are the combination of devices connected in shunt or in series. As depicted in Figure 8.2, devices connected in shunt are STATCOM and SVC, whereas devices connected in series are IPFC and SSSC.

Nowadays, second-generation power flow controllers are used and connected in series, but, along with controllers, inverters are also required to be connected with zero resistance so that voltage can be within limits, and conversion efficiency can be increased. FACTS controllers are classified on the basis of operation, series or parallel, duty cycle operation or conversion efficiency, and are:

- A thyristor controller, or
- A voltage source controller

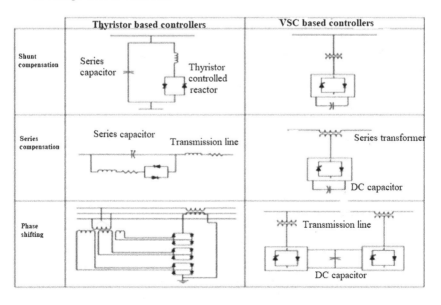

FIGURE 8.2 Summary of FACTS controllers [21]

Based on their methods of connection, the description of FACTS controllers is presented below:

- **Shunt controllers**: These controllers are of two types, namely variable impedance or variable magnitude, or a combination of both. Controllers connected in shunt inject power at the point of connection, e.g., SVC.
- **Series controllers**: Series-connected controllers are those controllers which inject power into transmission lines in series, e.g., SSSC.
- **Series-series combined controllers**: Combined series-series controllers are the combination of various series-connected FACTS controllers, e.g., two SSSC controllers in IPFC [22].
- **Shunt-series combined controllers**: Combined series-shunt connected controllers work as the combination of various controllers, e.g., in UPFC, one converter acts as a shunt controller (STATCOM),whereas the other converter works as a series controller(SSSC).

Typical characteristics of FACTS controllers are :

- Used for voltage control.
- Control the flow of reactive power.
- Used for damping power oscillations.
- Used for controlling imbalance conditions.

8.4.2 PRINCIPAL BENEFITS OF FACTS

The principal advantages of implementing FACTS controllers in transmission systems are :

- Analysis of load flow dynamics in a multi-machine system, keeping transmission line losses to a minimum. Due to this control of real and reactive power; it has been observed that losses increase to 3%.
- These controllers increase the utilization factor of a system, meaning that appropriate and optimal use of resources can be carried out. It also increases the ATC (available transfer capability) of the system so that the transmission line cost decreases, which, overall, reduces the total cost of electricity generation and reduces the burden on end-users.
- In the case of the transmission line, surge loading determines loading of the transmission line and decides its thermal insulation and voltage specification, which make a line suitable for a high-voltage line and reduces the cost of transmission lines.
- These devices provide security to the system and reduce interference with other lines subjected to other frequency limits.
- These devices enhance the stability of the system by largely reducing the surge losses and switching losses at the starting point or at the islanding point of connection.

- AFACTS controller reduces the requirement for reactive power (or compensation). It thus allows the transmission line to carry more real power.

8.5 UNIFIED POWER FLOW CONTROLLER (UPFC)

Figure 8.3 shows the configuration of a UPFC. A UPFC is a combination of series and shunt controllers in such a way that it controls real and reactive power control in a cascade manner. But both converters are connected through a capacitor so that the DC output should not be passed through a second converter and no circulating current should flow between the two converters. In other words, the UPFC controls voltage between elements of a transmission line in series configuration and maintains the voltage magnitude in accordance with phase and magnitude; while injecting reactive power into a system to allow uninterrupted flow of real power. Due to shunt capacitance, the voltage profile of a system is not maintained, and, as a result, voltage is dispersed, allowing a significant decrease in voltage. To avoid this kind of situation, UPFC connects an insulated gate bipolar transistor (IGBT) voltage source inverter (VSI) with zero resistance and operates it as a shunt controller, behaving as an earthing mechanism to keep the voltage drop to a minimum. These two converters have been connected with a common capacitor to block the DC supply to other converters. The rating of a capacitor is determined by surge loading of the transmission line and voltage specification of the line, along with the type of configuration. Additionally, UPFC is connected with the coupling transformer, with VSI as a series controller. The main objective of the shunt controller to keep the voltage constant across the capacitor, so that power loss is kept small, the total power available will be absorbed by a transmission line, and all losses will be equated and distributed among the series and shunt controllers of the UPFC [23, 24].

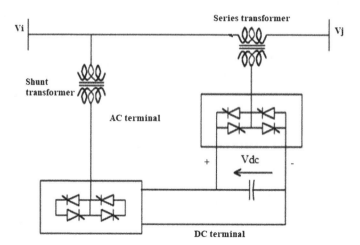

FIGURE 8.3 Configuration of a UPFC controller [25]

8.5.1 OPERATING MODES OF UPFC

It has already been discussed that, in UPFC, there are two types of converters: series and shunt. The shunt converter is operated to generate a controlled current into the line [26]. It is possible to control shunt converter by two modes, namely volt ampere reactive (VAR) control and automatic voltage control:

- **VAR control**: In UPFC, the reference voltage may be taken to be capacitive or inductive. Through VAR reference, the shunt current is established by the gate control of the inverter. In this control, feedback is of outmost importance and provided by DC voltage, due to fact that the DC signal feedback is of a step signal, with any deviation between desired output and actual output generating a trigger signal in terms of an error signal
- **Automatic voltage control**: Automatic voltage controls are divided into reactive current and series voltage. Reactive current (circulating current) operates across the shunt controller at PCC, whereas series voltage is controlled by injected reactive power into a transmission line to maintain phase angle and magnitude, with the symmetrical components remaining intact. The major concern is to stabilize the transmission with real and reactive power flow. These parameters can be analyzed and the operation can be summarized by the following modes:
  - **Direct voltage injection**: Reference inputs are given to UPFC (required and controlled) which are monitored on the basis of their magnitude and the phase difference between them.
  - **Phase-angle shifter emulation**: Deviation in input and output voltage keeps the reference voltage considerations intact.
  - **Line impedance emulation**: Matching of impedance is matched by the line impedance, to maximize surge loading
  - **Automatic power flow control**: Load flow dynamics, along with active and reactive power values, are used for maintaining the transmission line in reference modes

The UPFC is a second-generation FACTS device, which consists of two different converters, although the objective of both converters is to achieve load flow dynamics and to maintain real and reactive flow in a transmission line without losses in steady-state conditions. The UPFC is basically used for injecting power into the line; to maintain balance between the two converters, reactive power is absorbed and injected simultaneously in order to keep losses to a minimum, with zero harmonics in case of power electronic penetration. Despite the power electronic interface, both converters are thyristor controlled and have a triggering angle, therefore having unmatched switching capability between STATCOM and SSSC. Semiconductors behave as an insulator at room temperature, as a result of an increase in voltage; reactive demand is filled by the shunt converter, whereas the series converter provides voltage regulation, keeping input and output voltage at a constant frequency or at a frequency of voltage given to the transmission line. Zero regulation is also required to keep phase angle and magnitude intact, whereas the coupling transformer provides interrupted non-deviated voltage to a transmission line. Control of the voltage is directly monitored by a series converter in terms of real power, but, on the other hand, reactive power is being controlled in terms of absorption and generation by both converters. Likewise, the DC link provided by the capacitor is provided to exchange power flow between the two converters, keeping real power in synchronization. Both converters are self- sufficient in terms of absorbing and generating reactive power; this is a continuous process until balancing between real and reactive power is not maintained. In the steady-state condition, both converters keep the voltage profile constant and inject real power into the transmission line by controlling reactive power. In a faulty situation, such as a L-G fault, losses are increased and power flow decreased in terms of active power; often, in order to keep the system stable, voltage is injected into the system and reactive power is controlled, maintaining voltage control. Meantime, UPFC injects voltage into the transmission line and supports the system voltage. In this way, the system becomes stable and manages the flow of real and reactive power flow in a transmission line. By this process, the FACTS controller manages load flow dynamics, and stability is achieved.

8.5.2 Advantages and Disadvantages of UPFC

The principal merits of UPFC are:

- UPFC provides compensation for reactive power.
- Transmission line losses are reduced, improving the voltage profile.
- UPFC maintains power flow in the lines and voltage regulation.
- UPFC also improves voltage stability of the line.

Typical demerits of UPFC are:

- The overall reliability of the controller is affected by the presence of the capacitor.
- UPFC requires a more complex magnetic structure.

8.5.3 Role of UPFC in Reactive Power Compensation

Various controllers have been researched, along with heuristic algorithms to maintain stability of the system. Such controllers employ both converters in conjunction as series and shunt converters. The two converters absorb and generate reactive power, so that the series converters absorb reactive power if the line voltage is at 45 degrees with the line current, but where the angle is different from45 degrees, then the controller has the flexibility of either generating or absorbing reactive power as per requirements. The main objective is to keep the system stable by keeping the voltage interrupted on the transmission line, while keeping the voltage profile constant. Various case studies have proved that, with SSSC, series control provides greater efficiency for load dynamics as compared with second-generation controllers, being a first-generation controller, but IPFC and UPFC provide stability in accordance with power quality, power flow, and voltage profile of the system [27, 28].

Hybrid systems now dominate, taking over from conventional systems, so that a multi-variable approach is used, and, at the point of common coupling (PCC), power quality is a major concern, along with dynamic security assessment. It has been observed that voltage sag and swell, which occur due to deviation in voltage magnitude, relative to their reference values, produces harmonics which are dangerous in a system where multiple sources with different operating characteristics are connected. Due to advances in filter approach, UPQC is used as active filters, does not consume or take away energy from elements in the transmission line but supplies energy and is able to achieve voltage compensation in electric power systems [29]. Most researchers recommended UPFC and IPFC at most of the installations to reduce the overloading problem, as IPFC is connected at the multiline to achieve stability indices. Every FACTS controller must have a power electronic element as a major component, and produce harmonics, resulting in the generation of power losses at PCC. Due to these issues, a new term, known as power system stabilizer (PSS), has been incorporated into the feedback structure to control parameters of FACTS controllers, and mostly PSS has an inherent capability of controlling real and reactive power flow in a transmission line. IPFC is a different configuration of two series converters in which both converters are connected at two different lines and the DC link is provided to exchange power transfer. The main advantage is that both lines maintain the normal frequency, where any deviation in frequency would result in control of active power being totally dependent on the frequency value of the line [30, 31]. IPFC also maintains stability issues in accordance with switching and voltage drop across series and shunt capacitance.

Due to the decreasing use of fossil fuels and increasing environmental issues associated with these energy sources, renewable energy resources are being implemented on a large scale. Also, it has been seen that transmission line losses are the main issue in long transmission lines. To reduce such losses, voltage sags and harmonics, FACTS controllers play a very important role and control the real-reactive power of the system. In order to explain the power system stability of a double-fed induction generator, various types of WT generators, i.e., fixed-speed induction generator and double-fed induction generator, have been used to create electricity from wind power. Rotor winding is connected directly to the AC utility grid through a coupling transformer. It is connected to a converter and connected back-to-back with a voltage source converter [32].

8.6   OTHER AVAILABLE FACTS CONTROLLERS

A review of various FACTS controllers, normally used for management and stability of the flow of electric power, has been published [33]. The technology of FACTS is based on power electronic converters, which offer an opportunity to increase the controllability, stability, and power transfer capability of the transmission line. However, the inherent limitations of any transmission system to transfer stable electric power are angular stability, voltage magnitude, thermal limit, and transient and dynamic stability. Basically, FACTS controllers maintain voltage within acceptable limits. The main objectives of FACTS controllers are the loading of transmission lines near their thermal limits, regulation of electric power flow in transmission routes, and emergency control for prevention of cascading outages.

Controllers are mainly used to control real and reactive power of the system [34]. The complexity of the electrical power system has been increased, because of existing transformer load being near their closer limit. Analysis has also been carried out based on the changing nature of the power system. It has been seen that transmission lines are mostly fully loaded, which decreases the system stability [35]. Nowadays, thyristor-based phase-shifting transformers are also involved [36]. Due to short circuits, application of these transformers is limited. There are basically two types of thyristor-based series compensators in the system:

- Thyristor-switched series compensators
- Thyristor-controlled series converters

In order to achieve stability, multiple controllers are connected, in cascaded or coupled form and achieve a degree of freedom. These combinations of converters are thyristor controlled by changing the value of the trigger angle and changing the terminal voltage, while keeping the load current constant. However, all parameters remain constant in different configurations by the use of reactors (inductors) to absorb and generate reactive power by switching mode connected in series and parallel. In some of the controllers, due to magnetization and de-magnetization, the reactor burns up or loses control over load flow in maintaining real and reactive power within stipulated limits, with the increasing stability of the system and improved PQ.

8.7 CONCLUSION

This chapter has highlighted the importance of harnessing electric power generation through a wind power source, particularly a model of a wind power system capable of generating 5000 MVA. Overall, it has been found that wind power is available in varied form. In order to extract its maximum real power at all times, the MPPT technique is implemented in converter systems. Various controlling algorithms for MPPT techniques have been described. In order to address the PQ aspects of wind power plants, FACTS devices have been introduced and explained. Among the available FACTS devices, the need for and importance of UPFC has been explained. It has been found that UPFC is capable of managing real power in line and of improving PQ. This is due to the fact that UPFC is capable of quickly controlling power and injecting voltage into the line. It can also support the line voltage. The main advantage of UPFC is that it can quickly eliminate harmonics and feed a pure sine wave into the connected line. Finally, the chapter closes with the introduction of other available FACTS controllers, based on thyristor control and thyristor switching. Overall, FACTS devices have been found to be better suited to improving various PQ aspects of long-length transmission lines.

REFERENCES

1. N.G. Hingorani, "Introducing custom power", *IEEE Spectrum*, 32(6), pp. 41–48, 1995.
2. S.V.R. Kumar, and S.S. Nagaraju, "Simulation of D-STATCOM and DVR in power systems", *ARPN Journal of Engineering and Applied Sciences*, 2(3), pp. 7–13, 2007.
3. M.R. Azim, and M.A. Hoque, "A fuzzy logic based dynamic voltage restorer for voltage sag and swell mitigation for industrial induction motor loads", *International Journal of Computers and Applications*, 30(8), pp. 9–18, 2011.
4. B. Ferdi, C. Benachaiba, B. Berbaoui, and R. Dehini, "STATCOM DC-link fuzzy controller for power factor correction", *Journal of Acta Electrotechnica*, 52(4), pp. 173–178, 2011.
5. E. Acha, C.F. Esquivel, H.A. Perez, and C.A. Camacho, *FACTS: Modelling and Simulation in Power Networks*, John Wiley & Sons, Ltd., 2004, ISBN: 9780470852712.
6. F.A. Albasri, T.S. Sidhu, and R.K. Varma, "Performance comparison of distance protection schemes for shunt-FACTS compensated transmission lines", *IEEE Transactions on Power Delivery*, 22(4), pp. 2116–2125, 2007.
7. C.A. Camacho, "Phase domain modelling and simulation of large-scale power systems with VSC-based FACTS equipment", University of Glasgow, 2005, Ph.D. thesis.
8. V.P. Kumar, and A. Tamilselvi, "A control strategy for a variable speed wind turbine with a permanent-magnet synchronous generator", *International Journal of Modern Engineering Research*, 2(3), pp. 9–14, 2012.
9. P. Bawaney, and B. Sridhar, "Power distribution of wind diesel generator in isolated network", *International Journal of Scientific and Engineering Research*, 3(4), pp. 2–6, 2012.
10. J. Singh, and M. Ouhrouche, "MPPT control methods in wind energy conversion system". *Proceedings of the IEEE International Conference on Electrical Machines*, 16(5), pp. 321–329, 2011.
11. E. Koutroulis, and K. Kalaitzakis, "Design of maximum power point tracking system for wind energy conversion applications", *IEEE Transactions on Industrial Electronics*, 53(2), pp. 137–144, 2006.
12. H.G. Jeong, R.H. Seung, and K.B. Lee, "An improved maximum power point tracking method for wind power systems", *Energies*, pp. 1339–1354, 2012.
13. Y.S. Kim, Y. Chung, and S. Moon, "An analysis of variable speed wind turbine power control methods with fluctuating wind speed", *Proceedings of the IEEE International Conference on Electrical and Control Engineering*, 11(3), pp. 123–128, May 2013.
14. L. Shi, S. Dai, L. Yao, Y. Ni, and M. Bazargan, "Impact of wind farms of DFIG type on power system transient stability", *International Journal of Electromagnetic Analysis and Application*, 2(8), pp. 471–475, August 2007.
15. J. Zhang, Z. Yin, X. Xiao, and Y. Di, "Enhancement voltage stability of wind farm access to power grid by novel SVC", *IEEE Transactions on Power Delivery*, 2(1), pp. 2262–2266, April 2009.
16. Z. Zou, and K. Zhou, "Voltage stability of wind power grid integration". *Proceeding of the. IEEE International Conference on Electrical Machine System*, 3(1), pp. 1–5, August 2011.
17. B. Cevcher, T.E. Gums, S. Emiroglu, and E. Sahin, "PSO based determination of SVC control parameter for power system transient stability improvement", *Proceeding of the International Conference on Energy Conversion and Management*, 11(6), pp. 665–671, August 2009.
18. S. Kumar, M.K. Kirar, and G. Agnihotri, "Transient stability analysis of the IEEE-9 bus electric power system", *International Journal of Scientific Engineering and Technology*, 1(3), pp. 161–166, July 2012.

19. N. Shilpa, and S. Yadav, "Simulation of real and reactive power flow assessment with UPFC connected to a single/double transmission line", *International Journal of Advanced Research in Electrical, Electronics and Instrumentation Engineering*, 3(5), pp. 9–14, April 2014.

20. P.A. Desale, V.J. Dawdle, and R.M. Bandager, "Brief review paper on the custom power devices for power quality improvement", *International Journal of Electronic and Electrical Engineering*, 7(7), pp. 723–737, November 2014.

21. A. Singh, and B.S. Surjan, "Power quality improvement using FACTs devices", *International Journal of Engineering and Advanced Technology*, 3(2), pp. 132–135, December 2013.

22. V. Kumar, and S. Raju, "Application of interline power flow controller for power transmission system", *Electronics and Control Engineering*, 2(10), pp. 234–242, June 2014.

23. S. Akter, and P. Das, "Comparison of performance of IPFC and UPFC facts controller in power system", *International Journal of Computers and Applications*, 67(2), pp. 324–329, June 2012.

24. R.P. Singh, S.K. Bharadwaj, and R.K. Singh, "Flexible AC transmission system controller: A state of art", *International Journal of Electronic and Electrical Engineering*, 7(8), pp. 623–628, March 2014.

25. G. Nitin, and Y. Shashi, "Simulation of real and reactive power flow assessment with UPFC connected to a single/double transmission line", *International Journal of Advanced Research in Electrical, Electronics and Instrumentation Engineering*, 3(4), pp. 8384–8391, April 2014.

26. B. Singh, K.S. Verma, P. Mishra, R. Maheshwari, U. Shrivastava, and A. Baranwal, "Introduction to FACTs controllers: A technological literature survey", *International Journal of Automation and Power Engineering*, 1(9), pp. 145–151, March 2012.

27. P. Suman, P. Kumar, N. Vijaysimha, and C.B. Saravanan, "Static synchronous series compensator for series compensation of EHV lines Transmission lines", *International Journal of Advanced Research in Electrical, Electronics and Instrumentation Engineering*, 2(7), pp. 142–148, July 2013.

28. B. Singh, K.A. Haddad, and A. Chandra, "A review of active filter for power quality improvement", *IEEE Transactions on Industrial Electronics*, 45(5), pp. 960–970, October 2009.

29. R. Krishnan, and V.G. Krishnan, "Application of interline power flow controller as AGC", *Journal of Theoretical and Applied Information Technology*, 54(13), pp. 206–211, August 2013.

30. S.M. Ali, B.K. Prusty, M.K. Dash, and S.P. Mishra, "Role of FACTs devices in improving the power quality in grid connected renewable energy system", *Journal of Engineering Research and Studies*, 2(3), pp. 420–429, August 2013.

31. M.J. Kadhim, and D.S. Chavan, "Stability analysis of DFIG-based wind farms using different bus system", *International Journal of Innovative Science Engineering and Technology*, 2(4), pp. 248–256, March 2013.

32. A. Masoud, Q. Hassan, and A. Mahmud, "Flexible AC transmission system controllers: A review paper", *Proceedings of the Multi-Topic Conference on Wireless Networks, Information Processing and Systems*, pp. 843–850, January 2015, Jamshoro, Pakistan.

33. R.P. Singh, S.K. Bharadwaj, and R.K. Singh, "Flexible AC transmission system controllers: A state of art", *International Journal of Electronic and Electrical Engineering*, 7(8), pp. 234–242, March 2014.

34. M.J. Hossain, H.R. Pota, M.A. Mahmud, and R.A. Ramos, "Impact of large scale wind generators penetration on the voltage stability of power system", *IEEE Transactions on Power Delivery*, 6(1), pp. 1–8, July 2011.

35. S. Francis, S. Sreedharan, and S. Manakkal, "Voltage stability analysis of wind inte-grated grid", *Engineering Science and Technology, an International Journal*, 3(2), pp. 1–8, July 2011.

36. D.V. Hertem, and J. Verboomen, "Power flow controlling devices: An overview of their working principles and their principal range", *Proceedings of the IEEE 2005 International Conference on Future Power Systems*, November 2005, Netherlands.

9 Electric Energy Systems

Kamal Kant Sharma, Akhil Gupta, and Akhil Nigam

CONTENTS

9.1   INTRODUCTION

The needs of society and the dependence of gross domestic product (GDP) on reliable performance parameters for the production of electrical energy have increased exponentially. The main reason for this increase is the enormous use of electrically enabled devices and the "rat race" between developing and developed countries for a share of the profits from industry. The energy system is also making progress in the production of electricity by employing advanced technologies with a focused effort to make the system more reliable and sustainable. For more than a century, the electric power industry has developed across the whole world, with the introduction of different energy production sources [1]. There are basically two types of sources, namely conventional and non-conventional energy sources.

Conventional fossil fuel-based technologies, which have dominated electricity generation for decades have many drawbacks and various limitations, such as greenhouse gas emissions, pollution and the destruction of aquatic life, and the global impacts on climate change. Conventional power plants are installed in a centralized system and distribute electrical energy to different parts of the country, depending upon revenue and per capita consumption. The centralized structure enables the supply of electricity to sites which are far away from a generation site and requires a high-voltage insulated network. These conventional plants are suitable for large-scale distribution, but, for smaller scale generation and supply, distributed generation is used, known as a customized power plant [2–4]. These latter interconnected networks install various components like busbars, transformers, circuit breakers, and relays to transmit electricity efficiently. Sometimes, due to the presence of various losses, due to faults, there may be some loss of efficiency. CO_2 emission from fossil-fuel power plants also reduces the efficiency of electricity generation from power plants. Hence, these power plants need modern solid-state electronic devices or smart devices, together with advanced technologies, which play a crucial role in reducing these losses and which control environmental pollution, to achieve greater efficiency.

Various observations have been analyzed from current scenarios, considering a mismatch between demand and supply of electricity at the customer's site. There are many problems facing the growing power industry to meet the gap between the existing supply and distribution structure and what is required. Currently, the electrical power industry is facing some unprecedented challenges and opportunities. The industry is entering an era of significant change through the introduction of advanced technologies, electricity market changes, new consumer behavior, and regulation. The existing electric grid also needs to meet higher standards, such as with respect to reliability, security, efficiency, sustainability, and service cost.

More importantly, in the 21st century, societal needs have become very different from those observed a few years ago. Better quality, greater reliability, increase sustainability, and greater efficiency have all become basic consumer needs for improved electric power and storage systems. In traditional times, these were available in a less amount because of a scarcity of resources [5]. Maintenance cost is also an important issue for the electric energy system. Due to the presence of hazardous conditions and faults, utilities are facing novel challenges. These challenges may occur from the consumer side, which need to be addressed to maintain the repair and upgrading of the necessary infrastructure [6–10]. More advanced technologies and signal conditioning devices may improve these problems of overloading. These may also help to reduce maintenance costs and improve efficiency. All these difficult challenges require proper consideration from the regulatory authorities [11–14]. Consumers must be acknowledged by being provided with full information regarding the costs and benefits of facilities, and the availability of the most appropriate infrastructure.

9.2 CLASSIFICATION OF ENERGY SOURCES

Conventionally, there are two kinds of energy sources: primary and secondary. Primary energy sources are mostly suitable for direct use without any conversion from one form to another e.g. solar, wind, fossil fuel and natural resources [15–17].

These sources are directly used for generating electricity but are limited in use; in the case of solar and wind energy sources, conventional systems cannot be incorporated directly due to the absence of reactive power and other types of generators, such as solar plate collectors and doubly fed induction generator (DFIG). In consideration of solar and wind energy sources, output from solar is only in the form of DC, whereas output from wind is variable and fluctuating in nature. Therefore, fossil fuels are still widely used, but coal-fired power stations produces flue gases due to incomplete combustion of fossil fuel, and these represent environmental hazards.

Major secondary energy sources are hydrogen and synthetic fuels. Another classification of energy resources is based on the time required to regenerate an energy source, such as renewable energy source [18, 19]. Renewable energy sources are not depleted by human activities. Sources which are renewable by nature, depending upon climate, include tidal energy, geothermal energy, biomass, and wind, as well as rainwater harvesting and waste management. It has been reported that 25% of total energy reserves are renewable resources, compared with70% of the reserves being coal-based [20, 21]. On the other hand, fossil fuels are dependent upon various human activities and day-to-day management of waste products and greenhouse gases. Figure 9.1 show classification of various energy sources, which are normally employed to generate electric power.

Greater dependence on fossil fuels and greater control of human activities would be difficult to embrace in the current situation. On the other hand, renewable energy resources which can be customized and connected nearer to load, known as

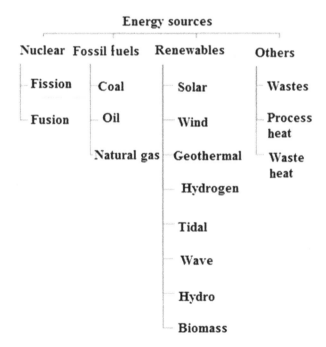

FIGURE 9.1 Classification of various energy resources [6]

distributed sources, could replace (at least in part) fossil fuels. Due to the increasing demand, distributed energy sources need to be connected with conventional sources but operational features are dissimilar from those of conventional sources, leading to fluctuation of voltage and changes in frequency. This kind of operational mechanism also leads to voltage instability and rotor angle instability under fixed conditions. Under dynamic operational conditions, practices of connecting conventional and distributed energy sources must be optimized with respect to environmental conditions, to ensure that all passive parameters obey maximum and minimum values of stability. These types of energy sources must be stabilized with compensation of the reactive power at the point of coupling in consideration of conventional and new sources of energy. This will improve the stability conditions and bridge the gap between load and supply with minimum disturbance [22].

9.3   COMPONENTS OF ELECTRIC ENERGY SYSTEM

There are various components and power conditioning devices employed in electricity generating power plants under different conditions. These components must be flexible and reliable to achieve proper functioning of the system during the generation of electric power. Basically, there are three major functions of electric power systems: electric power generation, its transmission, and ultimately its distribution to consumers.

- Generation terminology explains the cogeneration (non-conventional) and conventional generation principles and makes use of renewable energy sources, along with conventional generation sources, such as fossil fuel, hydro and nuclear. Renewable energy sources acts as peak power plants with a load factor of less than 100% unity, whereas conventional power plants act as base plants, with the exception of the hydro-power plant segregated as a run-off river, pumped-storage plant or a storage power plant without pumped storage, depending upon its capacity and requirements.
- The transmission of electric power involves the capacity of the transmission line to handle the value of voltage, which needs to be kept high. The higher the voltage of the transmission line and the greater the insulation, the lower would be the losses. Transmission lines are also segregated on the basis of different voltage levels and assessed on the basis of certain parameters, such as ATC (available transfer capability), contingency analysis, and conductors formation. Transmission lines can also be segregated on the types of voltage used, namely HVAC or HVDC.
- The distribution of electrical energy is crucial as the most losses occur in this part of the power system. Distribution is done at the level of area, and segregated as mesh, radial, and ring power systems, depending upon the load requirements. These distributions are circulated through transformers and feeders *via* substations, comprising various electrical devices for protection, and auxiliaries for smooth operation. Substations are evaluated on the basis of a voltage drop of ±5% and segregated as indoor or outdoor

substations. In hilly areas, H-pole mounted substations are used in order to compensate for the complexity in a system and the layout of power cables for distribution.

The most commonly used electric power generation network is shown in Figure 9.2. This depicts various components in which three-phase generation is accumulated by synchronous and non-synchronous generators. For protection, grounding terminology is used for individual components, whereas phase and line components are used for live connections considered neutral for safety (completion of circuit). Generation of voltage is done at multiples of 11, keeping sinusoidal waveform form factor 1.11 intact so that distortion and noise can be minimized, and regulated voltage can be obtained. The level of voltage from the generation point of view is variable and subjected to vary at different loads as the maximum voltage level of transmission is 765 kV and the optimum value of generation is 11 kV which generates minimum losses with maximum electrical capability of the system. The operational value of voltage increases with insulation level, consisting of various components, but mostly generation is accomplished by a three-wire system, having three phases, but, in the case of transmission, an extra conductor is provided at the top of the conductor, and is termed the earth conductor, and is required for safety. Remote generation is accomplished from distributed generation sources operating at the same level of utility voltage for achieving synchronization and being close to the customer site, and also known as cogeneration plants or stand-alone plants, if not connected with the main grid. Transmission capacity is governed by ATC or the maximum voltage-withstanding capacity of the line. Depending upon the type and level of voltage, layout considerations are implemented in the form of overhead or underground lines; underground lines incorporate cables whereas overhead systems require basic transmission layout at nominal voltage and current. The main constituent diversification of an electric power system is a distribution system which emphasizes voltage drop and feeder installation. In the case of a distribution system, the type of transformer decides the loading characteristics and efficiency of the system, and transformers used for step-down are segregated as core and shell transformers [21]. The segregation of voltage

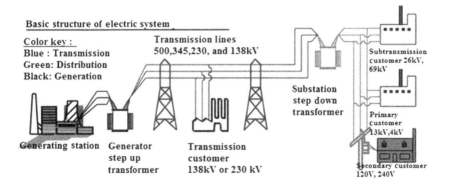

FIGURE 9.2 Schematic of a basic electric energy system

at the distribution end is limited to 400 V for industrial purposes and the agricultural sector, where load is intermediate and fluctuating in nature. Keeping voltage at a high value results in protection against electricity theft but safety considerations are of priority as consumers are involved in practices of distribution. Then, the operating voltage is stepped-down by using step-down transformers at a substation site to 132 kV, which is further stepped down to 34.5 kV for a normal distribution system. Eventually it is reduced to 230 V for domestic appliances and 400 V for industrial purposes [23].

9.3.1 Electric Power Plants

There are many industrial electric power plants, which play a major role in the generation and transmission of electric power. These employ sunlight ("solar photovoltaic"), wind, or hydro power sources which are converted into electricity through employing power converters, and which then transmit it. Sunlight is directly converted into electricity, whereas wind energy is converted into electric energy. There are two types of generation plants: conventional and non-conventional; Conventional plants involve fossil fuels, a resource available in abundance and ready to use for a long time period. As a consequence, these kind of plants can be located far from the consumer site and used in centralized power generation, whereas non-conventional (dispersed)generation involves intermediate energy generation, and supplements the conventional resource plant, requiring the assistance of converters and non-conventional generators, and having the disadvantages of reactive power compensation and voltage instability. These kinds of resources for electric power generation are segregated on the basis of the type of generation, the connected network, amount of generation, load factors, type of fuel used, and Kyoto Protocol indices. The main factor in dividing these resources into the two categories, based on the Kyoto Protocol, is to differentiate the economic constraints between developing and developed countries.

Figure 9.3 shows different energy conversion methods to generate electricity through different processes, using conventional or non-conventional energy sources.

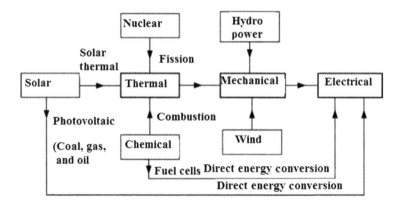

FIGURE 9.3   Different methods of electric energy conversion

The solar photovoltaic (PV) system directly converts sunlight into electricity; it does not need any alternative or intermediate process for energy conversion. In the case of wind power, different types of generators are used, like permanent magnet synchronous generator (PMSG), DFIG, self-excited induction generator (SEIG), using self-excitation and separate excitation, with the help of field winding; controlling reactive power changes at the point of coupling in an interconnected system causes fluctuation and disturbance. As a result, both solar and wind configurations of electric power systems for generation produce or trigger conversion from kinetic energy, or thermal energy into electrical energy, whereas, in the case of conventional systems, energy is converted directly into electricity without intermediate processes, and losses are minimal in consideration of the total power losses occurring with different configurations. Dispersed generation technologies use a renewable energy source in order to improve AFI and HUI and to reduce carbon footprints, leading to a decentralized power system in which the various components are assembled together for smooth functioning of a power system with limited constraints.

This type of energy conversion brings to light the fact that, during conversion from high value to low value, or *vice versa*, the system needs to be controlled from a stability point of view with respect to basic parameters like voltage, current, rotor angle stability, and dynamic stability of the generators. As per capita electricity consumption in Asian countries is increasing rapidly, and transformation from developing to developed country status risks breaching Kyoto Protocols, there is a need to increase use of dispersed generation sources, along with conventional sources, to maintain supply and demand proportionally. This kind of system transformation leads to a new kind of system, known as an embedded hybrid system, which brings different energy resources (conventional and non-conventional) together in an integrated hybrid approach for decision making with enhanced features. Nowadays; new terminology also applies to a system containing all components intact with old one, namely "Smart Grid", which is already in operation in some countries for improvement of the existing system, with fewer losses and greater communication interface. It is an amalgam of communication technologies, ICT, cyber security, and grid architecture with virtual machine space, along with heuristic algorithms for an intelligent framework. These kinds of systems involve self-optimization techniques with healing capacity at the occurrence of a fault, with physical machine conversion to a virtual machine, using SVM (support vector machine),and constraints of physical parameters. This kind of transformation represents a new beginning for electric power systems, with more intelligent techniques in use in the network, with full connectivity to storage, security, and reducing the risk of manpower malfunction for coming generations.

9.3.2 Grid Modeling Systems

Grid modeling needs the more advanced control approaches as modernization progresses from the conventional energy resources to non-conventional ones. As discussed in Section 9.3.1, the new system emerges as a Smart Grid (Intelligent Grid),consisting of basic components of the existing grid for the physical space, but virtualization of physical machines into their respective vector components leads to a different architecture, segregated into cluster, cloud, or database management system (DBMS) servers. These types of changes enable the user to accommodate their usage amid real-time implementation of the sources, and they can be accessed by different GENCO (generation companies) to overcome the surge usage and control the same by changing the dynamic parameters of the system. This will save the resources available, and make optimized use of resources allocated as per the agreed scheduling of load among the consumers [24]. Smart Grid makes use of cyber space, fuzzy and crisp algorithms with decision-making technologies on a real-time basis, equipped with protection systems against severe faults. In addition to this, certain modeling features could be extracted with different grid codes and technology transfer on a common interface, using vitalizing algorithms for better communication and interface among them. Figure 9.4 depicts the development of fuels in the future, and how these energy sources can be used. Several projections regarding various energy systems are summarized below:

- Energy consumption will be doubled by 2050, with renewable energy sources making up over 50% of electricity generation by the year 2035.
- Gas usage will continue to increase its share of supplying global energy demand.
- Oil demand growth will slow down substantially, with peak time in 2030.
- Carbon emissions will decline, due to decreasing coal demand, by 2050.

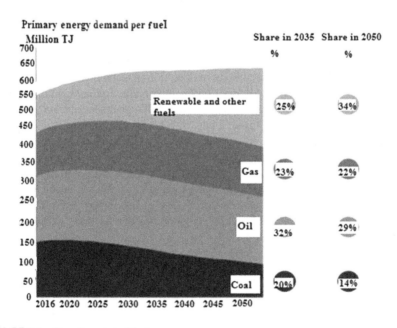

FIGURE 9.4 Development of fuels

Nowadays, electricity share has increased considerably to 10 times GDP, utilizing every energy resource for reducing the gap between demand and supply. Electricity is a commodity which is needed by everyone and on which every task of humankind is dependent. In European countries and South America, decentralized power systems exist, with the sale and purchase of electricity allowed on an individual basis without government intervention. In India, initiatives have allowed Government agencies to increase stakeholders among private companies, and power exchange has also been established, with a need to introduce competitiveness among various players to provide electrification to rural India on a priority basis. Advances have been incorporated into the agricultural sector and modernization of economies in the rural sector allow private companies to bring resources to the customer at relatively low prices and to improve tariff charges to the benefit of customers. The use of generators is also useful and can be turned on and off on a real-time basis, so that generation can be achieved appropriately at the generator side without loss of productivity, keeping the load indices constant. These technologies represent a path-breaking move by the government to bring change to the existing system and to move toward a sustainable future for generations to come.

9.3.3 Energy Storage Systems

Many years ago, electricity could not be stored properly, a shortcoming which encouraged its immediate consumption. Nowadays, saving of electricity leads to more generation of electricity. In spite of this, the significance of saving electricity is regarded at three on a scale of one to ten, and is not considered to be a top priority, but various technologies have overcome these problems and provided sufficient interface by which to store adequate amounts of energy for users in remote locations or for emergency situations. These types of solutions are becoming popular with existing consumer but are still out of reach of a remote consumer or who cannot afford these solutions, which are not available at remote locations or are available at high prices in sub-urban areas. Energy generation is increasing in response to consumer demand, and energy storage is essential to store energy for its utilization, especially when the energy source is out of service due to a fault. Hence, there are major challenges for us to store electric energy. Developments has been made in this sophisticated area but it is feasible only to grid-connected stations, major stakeholders (in terms of per capita consumption), government-owned GENCO companies or to cater for large industrialists. Research into expanded storage strategies is very limited, because of the non-profitability of such an approach. It has been observed that storage devices for electric power have very few applications, are limited to a specific range and cannot be customized, despite the range of devices available. Over the past decade, some experimental work on wireless electricity has been carried out, although this still seems a distant dream. Therefore, very few sectors require energy storage, whereas some of the techniques, like islanding, are viable and are used worldwide to assess the severity of faults on electric power systems.

There are many possible techniques for energy storage for all forms of energy, like thermal, chemical, and mechanical, as depicted in Figure 9.5. The storage techniques can be classified into four categories, depending on their applications:

- Low-power application feed transducers and emergency terminals.
- Medium-power application in isolated areas, such as individual electrical systems and town supplies.
- Network connection application related to peak leveling.
- Power-quality control-related applications.

Storage technologies can be distinguished in terms of star rating, duty cycle, state errors, technology type used, thermal sinks, system size incorporated, compatibility with the existing system, range of equipment needed, insulation level, and interface technologies available. Storage systems can be available as an amalgam of different technologies, or as various branches of engineering or reactions through various processes, or as by-products of an individual component. Principal component analysis (PCA) can be conducted to evaluate performance analysis of storage systems, subject to constraints available with existing systems. Energy storage technologies can be depicted as shown in Figure 9.5. Figure 9.5 also shows the different processes associated with certain branches of engineering and their application-based technologies. Similarly, in mechanical storage, the use of flywheels dominates, along with pumped-storage systems with controlled air pressure, which can be used *via* nozzles or a combination of low-pressure to high-pressure systems, and *vice versa*. In this kind of storage classification, under the cluster of "mechanical storage system", kinetic or potential energy (dynamic or static energy) is stored, whereas, on the other

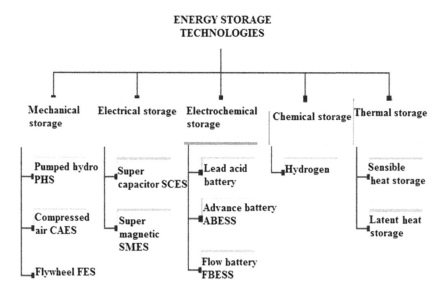

FIGURE 9.5 Classification of energy storage technologies [23]

hand, the flywheel is used to store energy when the system changes its state from ON position to OFF position.

9.3.3.1 Pumped Hydro Storage

This kind of storage system comes under the category of mechanical storage system and is used in power plants consisting of hydro-energy resources. In the reservoir, whenever an amount of water is required for electricity generation, then water is pumped from low to high elevation, with the help of a motor or some electrome-chanical device. The pumped system is incorporated into hydro-power plants and used in the case of a run-off river power plant or a run-off river plant without pond-age, in which sufficient amount of water is not available in off-peak times or seasons. It also acts as a surge tank whenever pressure is not maintained but released over a period of time, affecting electrical equipment with a flow of water, losing insulation. During the period of water requirement, water is released for the turbine, electric-ity is generated, and a load-balancing mechanism is performed. During off-peak and peak periods, the difference in the amount of water is significant and the use of pumped storage is required to sustain electricity generation at a time when water is not available. It is a mechanical auxiliary system, maintaining the flow of water, as either turbulent or streamline flow, depending upon the rating of turbine in terms of power and energy, and considering motor-generator sets. This kind of mechanism involves readily available equipment and its efficiency can be increased with certain tested mechanisms. A concentrated water supply is required, and it acts like a cata-lyst to increase productivity at an efficiency of greater than 60%.

9.3.3.2 Compressed Air Storage

Compressed air energy storage (CAS) basically utilizes air pressure in different states. It works like a nozzle or turbine structure; in the case of a nozzle, pressure is high but at the outgoing end, pressure is released, whereas in the case of a turbine, pressure is low at the input of the turbine and is high at the outset of the turbine. This whole process reflects changes in air pressure, but, instead of releasing pressure, compression is the best-suited mechanism for an energy source. It also acts like an off-peak mechanism, which stores energy during the off-peak period and releases it whenever demand is high. It is a mechanical system with frictionless intermedi-ate at a scale of 5 out of 10, interfaced with advanced technologies for high power requirements.

9.3.3.3 Flywheel Storage System

This kind of system can be likened to a ball-flying mechanism, such that, whenever a ball is released up into the air, it releases energy under the effect of gravity and comes back to earth. Similarly, a flywheel works by conservation of energy, which states that "Energy can neither be created, nor can it be destroyed". By applying the conservation law, a flywheel can be static or rotating in nature, as in power electronics circuits. A diode is used as the flywheel and operated in freewheeling diode mode, possessing different characteristics in forward- and reverse-biased conditions. Although a fly-wheel, by its rotating nature, accrues energy at a high relative speed and releases it at a low speed, it maintains stability and follows the equal area criterion. The flywheel is connected to a rotor, with the type of material (principally, the weight of material) determining the number of rotations. Minimum and maximum limits for the speed of the flywheel lies within the range of 10,000–1,00000 rpm. The value of speed depends upon the accelerating and decelerating powers of the machine.

9.3.3.4   Electrical Storage

This is the most efficient way to store electricity, as no conversion is required and no intermediate component for either coupling or cascading is needed, but the main roadblock for storing electrical energy directly is the lack of availability of the materials for such applications. One of the technologies used in the storage of electrical energy is super capacitors and super magnetic electrical storage (SMES), but they are not readily feasible and are not available on a large scale.

9.3.3.5   Super-capacitor Energy Storage

In case of a super capacitor, a limit is imposed on the number of plates and the type of dielectric used. In the case of the capacitor, it is voltage controlled and must be accessed by an electric field. These devices have characteristics of capacitors with applied DC voltage, with no chemical reaction. They are not easy to fabricate, with a limitation imposed by the need for insulation levels.

9.3.3.6   Super-conducting Magnetic Energy Storage (SMES)

It has been proved experimentally that no insulator exists in magnetic circuits, although magnetic lines of force flow between the two poles of a magnet, but, in the case of SMES, the current flows through a super-conducting coil and heating is cooled down to a level where the temperature is less than the critical temperature, without changing the thermal sink. The main advantage lies in the fact that an inductive coil produces transient current at $t = 0+$ and $t = 0-$, forming closed and open circuits, respectively, and storing energy in a form of magnetic field.

9.3.3.7   Electrochemical Energy Storage

This terminology follows the traditional approach to storing electrical energy but depends upon a chemical reaction in the presence of an electrolyte. The main drawback is that they store electric energy in the form of chemical energy and follows a reaction consisting of only DC type electric energy. Another roadblock in this field is the selection of the materials used, particularly their melting point and their connection when dissimilar elements are connected together. There are various possibilities available, such as lead-acid, nickel-cadmium, nickel-metal hydride, nickel-iron, zinc-air, iron-air, lithium-ion, and lithium-polymer, with specific characteristics, ranging from certain levels of insulation to temperature rankings, of use for particular applications.

9.3.3.8   Batteries

Batteries are also used to store electricity and act as a primary storage system. The most common type of battery used is the lead-acid battery, which make use of lead dioxide (PbO_2) as the positive electrode and spongy lead (Pb) as the negative electrode, with sulfuric acid (H_2SO_4) as the electrolyte, with lead as the current collector. The main problem is selection of the material, as lead is toxic in nature and susceptible toa fixed range of temperature, so that the number of plates is also fixed (considering positive and negative plates). Lead-acid batteries are fast in response, with low cost of manufacturing and no leakage at the optimum temperature level. In addition to this, the battery energy system is also valuable in consideration of safety and capacity to work at positive and negative temperatures, ranging from −20°C to +40°C. The maximum allowable limit is +40°C. Various possibilities in selection of materials are nickel chemistry, sodium chemistry, and lithium chemistry. Lithium materials are quite expensive, and sodium forms alkaline base forms, which result in sludge formation at high temperatures and are hygroscopic in nature. New terminologies also come into the system, with terms such as redox flow batteries. These types of batteries use reduced oxidation levels of electrolyte in a battery with a true value of composite material.

9.3.3.9   Chemical Storage

Chemical storage is a new class of vertical storage devices, which is achieved using catalysts, various accumulators, and a combination of batteries in series and in parallel. It performs dual functions of storing energy and keeping the charge-discharge plates charged by providing current to the plates, so that energy is not wasted, but being used optimally. Major issues with chemical storage are the emission of harmful gases, excessive noise levels, which can cause tremor in human brains, so they are not used in human interface machines, and sensitive to exposure to extreme temperatures.

9.3.3.10   Thermal Storage

This type of storage system is the most tested one and exploits various principles, like the Seeback effect, the Peltier effect, and the Thomson effect. It has been segregated into high- and low-temperature systems, selected on the basis of their room temperature level in consideration with operating temperature. In addition to this, the storage medium responds to a change in temperature, with phase change, with an increase in high temperature, with a score of 7 out of 10. The storage medium can be a dielectric or insulating material, depending upon the electrical characteristics. In sensible heat storage, heat exchange exhibits phase change, using organic and inorganic materials with no temperature change, and is referred to as hidden energy. The latent heat storage system has a high storage density and an ability to store energy with small temperature variation.

9.4   CONCLUSION

This chapter has mainly addressed and highlighted the various resources of renewable energy resources which are capable to replace the conventional resources in modern grids infrastructure. It has also highlighted the need with the increasing trend towards increasing loads of all categories. Thus it has been concluded that that these energy systems can play a significant role in the developments of modern grids and associated control systems.

REFERENCES

1. T. Gonen, *Electric power distribution engineering*, 3rd ed, Florida: CRC Press; 2014.
2. N. Mohan, *Electric power system first course*, New Jersey: John Wiley & Sons; 2012.
3. G.M. Masters, *Renewable and efficient electric power system*, New Jersey: John Wiley & Sons; 2010.
4. F. Wu, K. Moslehi, and A. Bose, "Power system control centers: Past, present and future", *Proceedings of the IEEE*, 93(11), 2005, pp. 1890–1908.
5. S. Chapman, *Electric machinery and power system fundamentals*, New York: Mc-Graw-Hill; 2001.
6. I. Dincer, and A.A. Rayash, *Energy sustainability*, United Kingdom: Elsevier, pp. 19–58; 2020.
7. J.J. Grainger, and W.D. Stevenson, *Power system analysis*, New Jersey: Mc-Graw-Hill; 1994.
8. L. Freris, and D. Infield, *Renewable energy in power systems*, West Sussex, UK: John Wiley & Sons; 2008.
9. Facts and figures, Energy storage association, 2013 Available from: http://energystorage.org/energy-storage/facts-figures.
10. F. Beguin, and E. Frackowaik, *Carbons for electrochemical energy storage and conversion systems*, Boca Raton: CRC Press Taylor and Francis Group; 2010.
11. B.E. Conway, V. Birss, and J. Wojtowice, "The role and utilization of pseudo capacitance for energy storage by supercapacitors", *Journal of Power Sources*, 66(1–2), 1997, pp. 1–14.
12. J.I. San Martin, I. Zamora, J.J. San Martin, V. Aperribay, and P. Eguia, "Energy storage technologies for electric applications". *International Conference on Renewable Energies and Power Quality*, Las Palmas, Spain, 2011.
13. International Standard Norme Internationale, IEC standard voltages, Horizontal Standard IEC 60038, ed. 7.0, 2009, pp. 1–11.
14. L. Stephan, "Mapping the future of energy storage in European energy markets", *Energy Storage Update Conference*, Younicos, November 2015.
15. M.A. Hannan, F.A. Azidin, and A. Mohamed, "Hybrid electric vehicles and their challenges: A review", *Renewable and Sustainable Energy Reviews*, 29, 2014, pp. 135–150.
16. X. Luo, J. Wang, M. Dooner, and J. Clarke, "Overview of current development in electrical energy storage technologies and the applications potential in power system operation", *Applied Energy*, 137, 2015, pp. 511–536.
17. M. Macia, D. Grabowski, M. Pasko, and M. Lewandowski, "Compensation based on active power filters: The cost minimization", *Applied Mathematics and Computation*, 267, 2015, pp. 648–654.
18. H. Chen, T.N. Cong, W. Yang, C. Tan, Y. Li, and Y. Ding, "Progress in electrical energy storage system: A critical review", *Progress in Natural Science*, 19(3), 2009, pp. 291–312.
19. B. Buchli, D. Aschwanden, and J. Beutel, "Battery state-of-charge approximation for energy harvesting embedded systems, wireless sensor networks". *Lecture Notes in Computer Science*, 7772, 2013, pp. 179–196.
20. L. Yuan, X.-H. Lu, X. Xiao, T. Zhai, J.Dai, F. Zhang, B. Hu, X.Wang, L. Gong, J. Chen, C. Hu, Y. Tong, J. Zhou, and Z.L. Wang, "Flexible solid-state supercapacitors based on carbon nanoparticles/MnO2 nanorods hybrid structure", *ACS Nano*, 6(1), 2012, pp. 656–661.
21. M.G.Y. Batarseh, "Components of electric energy systems", *Electric Renewable Energy Systems*, London: Academic Press, pp. 21–39; 2016.

22. L.R. Martinez, and N. Omar, *Emerging nanotechnologies in rechargeable energy storage systems, 1st edition*, Spain: Elsevier, p. 346; 2017.

23. B.K. Bose, *Power electronics in renewable energy systems and smart grid: Technology and applications*, New Jersey: John Wiley & Sons, Inc., p. 752; 2019.

24. A.A. Sallam, and O.P. Malik, "Electrical energy storage" *The Institute of Electrical and Electronics Engineers, Inc.*, New Jersey, John Wiley & Sons, Inc., pp. 535–552; 2019.

10 Power Quality Analysis Using Various Computational Techniques for Induction Machines

Akhil Gupta and Sunny Vig

CONTENTS

10.1 INTRODUCTION

The current deregulated industry is mainly responsible for the introduction of Power Quality (PQ) issues in the electric power system; these issues, which are quite challenging in the long run, have been major concerns in recent times. Therefore, the impact of these issues needs immediate consideration on account of the sudden increase in harmonic-producing loads. However, the existence of harmonics is inevitable at the user end [1–3]. Critical sources of harmonics are rotating machines, inverter drives, non-linearity of a transformer, phase controllers, and AC regulators. In particularly, the presence of any level of harmonic in the system depends upon the switching pattern employed. That is why a novel hybrid phase shift Pulse Width Modulation (PWM)-modified synchronous optimal hybrid techniques has been proposed [4] for a 27-level multi-level inverter, which was aimed at improving the harmonic voltage profile of a 4 kW solar photovoltaic grid-connected system.

There are numerous categories of drive systems being employed in several applications, such as electronic devices [5], in the control of electrical vehicles [6], as well as in wind energy-conversion systems [7]. It has been reported that motors typically adopted for these applications are three-phase Induction Motors (TIMs), with a squirrel-cage rotor, due to their widely known advantages, such as their simple construction, ruggedness, and low cost [8]. Combinations of all these features makes these motors ideal in various speed-regulating and brushless operations in the healthcare sectors. In addition, medical product suppliers and pharmaceutical organizations can implement their product concept using these drive systems, which consistently focus on the patients' needs. As a result, drive technologies and control strategies started to be developed and applied to TIM systems [9].

Basically, the scalar (V/f) control, the Field-Oriented Control (FOC), and the Direct Torque Control (DTC) are the methods frequently adopted to drive TIMs [10]. A switching strategy for FOC is proposed for minimizing the power of filtered errors, by modifying the operational frequency and the pulse-width resolution, which reduces the switching number without sacrificing speed performance [11]. AFOC-based nonlinear tracking control has been proposed, together with a new adaptive local flux observer [12]. For a TIM, an experimental framework for tuning of proportional integral controllers of a DTC drive, controlled by space vector PWM, has been reported [13]. Ripples in electromagnetic torque and current harmonics have been drastically reduced by 69.2% and 47.7%, respectively, for a doubly fed induction machine [14]. Fuzzy-level blocks have been employed in DTC strategies, instead of hysteresis regulators and voltage vectors. Therefore, it has been found that modulation techniques play a vital role in PQ control and the synthesis of balanced output voltages and currents. Especially in TIMs, the proposed modulation techniques in current research eliminate the triangular carrier waveforms and implementation complexities. The scope of this research is to make medical professionals, who work in patient monitoring, with respect to the conduct of postoperative therapies, irradiation techniques, and wound management, aware of the PQ aspects of TIM drives.

This chapter mainly presents the harmonic mitigation impact of two switching techniques, namely PWM and Discrete Space Vector PWM (DSVPWM), caused by connecting a 3-HP TIM. Insulated Gate Biplor Transistor (IGBT)-based rectifier and inverter systems have been used where PQ assessment has been presented under different loading condition patterns. In particular, the impact of DC-offset, 3rd and 5th harmonics has been alleviated by an appropriately designed Inductance-Capacitance (LC) passive filter. Through an experimental set-up, harmonics have been measured at various points of common coupling through a MECO-based Power and Harmonics Analyzer-PHA 5850. The superiority and effectiveness of the DSVPWM technique over Sine PWM were established, in obtaining balanced output voltage and current waveforms at reduced DC-offset and harmonics levels. Measurements at various Point of Common Couplings (PCC), using the Power and Harmonic Analyzer (PHA), have demonstrated the effectiveness of proposed modulation switching techniques. Thus, PQ estimation ensures the applications of TIM drives especially in medical applications, where reliability and diagnostic precision are the top priorities.

10.2   COMPUTATIONAL MODELS

As depicted in Figs. 10.1 (a) and 10.1(b), the rectifier-inverter model was controlled by modulation techniques for generating a controlled switching signal. This signal has been generated through two types of modulation techniques. Basically, the proposed system is composed of a AC-DC-AC insulated-gate bipolar transistor (IGBT)-based converter, which connects a three-phase supply to TIM. The inverter used for DC-AC conversion is operated using the PWM or DSVPWM modulation technique. Thus, by designing a passive filter, the objective of improving the PQ is achieved. PWM is a switching technique, which works efficiently at the higher

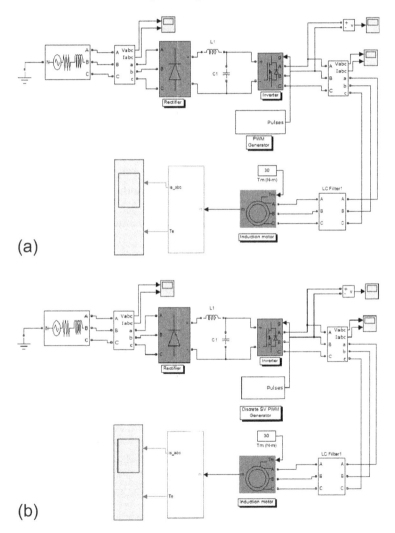

FIGURE 10.1   Simulink model of AC-DC-AC conversion system using a) PWM and b) DSVPWM modulation

carrier frequency [15–17]. In this technique, the duty ratio of a sinusoidal pulsating waveform is controlled with the help of a triangular-type waveform. A controlled switching signal is generated through the intersection of a reference sinusoidal voltage waveform and a carrier waveform, which controls the operation of IGBT-based switches and the speed of the electric drives. The longer the duty ratio of switches, the higher is the active power delivered to the connected load. The switching frequency of the discrete SVPWM switching technique is 1080 Hz. The proposed system is capable of controlling PQ by alleviating 3rd and 5th harmonics generated in the presence of a three-phase SQIM of 3 HP, 4.5 A, 410 V, 50 Hz-rated current as a connected load [18].

Normally, the three-phase supply consists of harmonics of varying levels. To mitigate the impact of harmonics, AC supply is first converted into DC supply through the rectifier and then reconverted into AC supply through the inverter. By generating a controlled switching signal, the AC-DC-AC converter has been coordinated to reduce the harmonics.

Fig. 10.2 illustrates the block diagram of a proposed model of an AC-DC-AC conversion system, using two PWM modulation techniques. At the input side, a three-phase supply has been connected with a three-phase rectifier system to synthesize DC power. A *LC* filter has been designed and connected for removing higher-order harmonics and high-frequency transients. A passive filter makes the waveform smoother and reduces the remaining ripple content of the line current [19, 20]. The converter acting as an inverter is coordinating with either the DSVPWM or PWM switching technique.

With the emerging technology, the SVPWM technique has been playing a pivotal role in power conversion and reducing harmonics. DSVPWM is a modulation algorithm, which translates phase voltage reference into modulation times. Instead of maintaining the width of all pulses as being similar in the case of multiple pulse width modulation, the width of each pulse is varied in proportion to the amplitude of a sine wave evaluated at the center of a similar pulse. By doing this, the distortion factor and lower-order harmonics are reduced significantly. The gating signals are generated by comparing a sinusoidal reference signal with a triangular carrier wave of frequency. Several of the important superiorities of the DSVPWM technique over the SPWM technique are a wide linear modulation range, lesser switching loss, less

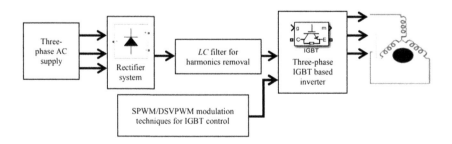

FIGURE 10.2 Block diagram of AC-DC-AC PWM-controlled converter system

total harmonic distortion (THD) in the spectrum of the switching waveform, ease of implementation, and fewer computational calculations required [21].

10.3 MATHEMATICAL ANALYSIS

Importantly, harmonics are present in power systems due to regular usage of non-linear loads. The presence of harmonics causes increased core losses and skin effects, reduction in applied torque, damaging of internal insulation, increased electromagnetic interference, and the resultant deviation in torque-speed curves. The harmonics present in the induction motor are time and space harmonics. The source may contain odd harmonics such as 3rd, 5th and 7th harmonics. Even-ordered harmonics cannot be present due to the symmetry of the waveform as $f(t) = -f(t+T/2)$. Let the r phase of the machine be given by:

$$V_R = V_{1m}\sin(\omega_1 t + \varphi_1) + V_{3m}\sin(3\omega_3 t + \varphi_3) + V_{5m}\sin(5\omega_5 t + \varphi_5) + V_{7m}\sin(\omega_7 t + \varphi_7) \quad (10.1)$$

Assuming that it is a balanced three-phase supply so V_Y and V_B are 120° and 240°, respectively, shifted from V_R, then the expression for V_Y and V_B can be expressed by Equations (10.2) and (10.3):

$$V_Y = V_{1m}\sin\left(\omega_1 t + \varphi_1 - \frac{2\pi}{3}\right) + V_{3m}\sin\left(3\omega_3 t + \varphi_3 - 3*\frac{2\pi}{3}\right)$$
$$+ V_{5m}\sin\left(5\omega_5 t + \varphi_5 - 5*\frac{2\pi}{3}\right) + V_{7m}\sin\left(\omega_7 t + \varphi_7 - 7*\frac{2\pi}{3}\right) \quad (10.2)$$

$$V_B = V_{1m}\sin\left(\omega_1 t + \varphi_1 - \frac{4\pi}{3}\right) + V_{3m}\sin\left(3\omega_3 t + \varphi_3 - 3*\frac{4\pi}{3}\right)$$
$$+ V_{5m}\sin\left(5\omega_5 t + \varphi_5 - 5*\frac{4\pi}{3}\right) + V_{7m}\sin\left(\omega_7 t + \varphi_7 - 7*\frac{4\pi}{3}\right) \quad (10.3)$$

Now, if the 3rd harmonic component of the three-phase waveforms is considered, and if $Vx3\ (t)$ is the 3rd harmonic of phase x, it can be written as:

$$V_{R3} = V_{3m}\sin(3\omega_1 t + \varphi_3)$$

$$V_{Y3} = V_{3m}\sin(3\omega_1 t + \varphi_3)$$

$$V_{B3} = V_{3m}\sin(3\omega_1 t + \varphi_3) \quad (10.4)$$

Therefore, from Equation (10.4), it can be revealed that 3rd harmonic components are in phase, which does not cause the flow of current in a star-connected system. However, if a neutral is connected to the system, there will be a current flow which occurs occasionally in any induction machine. Thus, TIM operates in open-circuit conditions to 3rd harmonics, called triplen harmonics. The equations for the 5th harmonic can be expressed as:

$$V_{R5} = V_{5m} \sin(5\omega_1 t + \varphi_5)$$

$$V_{Y5} = V_{5m} \sin\left(5\omega_1 t + \varphi_5 - 5*\frac{2\pi}{3}\right)$$

$$= V_{5m} \sin\left(5\omega_1 t + \varphi_5 - 5*\frac{4\pi}{3}\right)$$

$$V_{B5} = V_{5m} \sin\left(5\omega_1 t + \varphi_5 - 5*\frac{4\pi}{3}\right)$$

$$= V_{5m} \sin\left(5\omega_1 t + \varphi_5 - \frac{2\pi}{3}\right) \qquad (10.5)$$

From Equation (10.5), it is concluded that the 5th harmonic component forms a negative sequence, which causes a backward-revolving flux pattern as compared with the fundamental frequency at their respective synchronous speeds. For the 7th harmonic component, the equations are expressed as:

$$V_{R7} = V_{5m} \sin(7\omega_1 t + \varphi_5)$$

$$V_{Y7} = V_{5m} \sin\left(7\omega_1 t + \varphi_5 - 7*\frac{2\pi}{3}\right)$$

$$= V_{5m} \sin\left(7\omega_1 t + \varphi_5 - 5*\frac{2\pi}{3}\right)$$

$$V_{B7} = V_{5m} \sin\left(7\omega_1 t + \varphi_5 - 7*\frac{4\pi}{3}\right)$$

$$= V_{5m} \sin\left(5\omega_1 t + \varphi_5 - \frac{4\pi}{3}\right) \qquad (10.6)$$

From Equations (10.6), it is concluded that the 7th harmonic component forms a positive sequence, which causes a forward-revolving flux pattern in order that the torque produced will be additive to the fundamental torque. Space harmonics are due to non-sinusoidal distribution of coils in machine and slotting. Practically, air gap magnetomorphic force (mmf) and flux are not sinusoidally distributed in space.

10.4   DESIGN OF PASSIVE FILTER

The passive shunt filter is the filter most commonly used to eliminate current harmonic distortion. In particular, these filters are connected in shunt with a connected load. Fig. 10.3 (a) depicts its electric circuit diagram, whereas Figure 10.3 (b) depicts its hardware setup. An ultimate objective is to mitigate harmonics, produced by non-linear loads, especially by induction motor drives. Although passive filters offer a low-impedance path to divert all harmonic components of current, they also have the tendency to offer some reactive power to the system. Importantly, the filter is used to serve two objectives: one for filtering purpose and the other for reactive compensation, which ultimately improves the power factor of load circuit. Therefore, the passive shunt *LC* filter is designed which mitigates 3rd and 5th order harmonics and reduces the overall THD of the line current-voltages. Chosen parameters of the inductor and capacitor of the shunt passive filter are 6 mH and 6.5 mF, respectively.

10.5 SIMULATION RESULTS, ANALYSIS, AND DISCUSSION

In this proposed work, the PHA-5850 was used for carrying out PQ analysis at various PCCs. MATLAB simulations were performed and validation was achieved by developing an experimental set-up. For measurements, the PHA was connected in between an autotransformer and a direct online (DOL) starter, using current probes and voltage probes for evaluating harmonics in line current and line voltage, respectively. This investigation was carried out under varying loads on a squirrel cage induction motor (SQIM), namely no-load, half-load, full-load, and under blocked-rotor. Fig. 10.4 (a) illustrates the block diagram of an experimental setup of an SQIM with the PHA. Rating of a three-phase SQIM is 3-HP, 4.5 A, rated voltage 415 V,

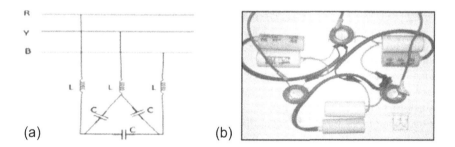

FIGURE 10.3 Typical passive filter. (a) Electric circuit diagram, and (b) hardware set-up

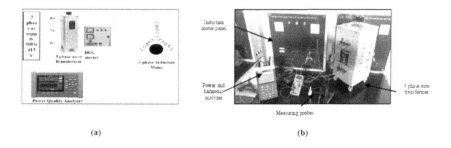

FIGURE 10.4 (a) Experimental setup with PHA (b) Experimental setup of TIM with PHA

rated speed 1440 rpm and B class insulation. Furthermore, an autotransformer is connected, which is used for varying the supply voltage. The DOL starter is used for starting the three-phase SQIM. Fig. 10.4 (b) depicts the complete hardware setup for a proposed scheme in which red, yellow, and blue probes have been used for measuring harmonics in line current. Similarly, red, yellow, and blue clamps are used for measuring harmonics in line voltage. Current measurement range is 0.1 mA to 1000 A. Also, the PHA is capable measuring harmonics upto the 99th order, with a display of 50 harmonics displayed on one screen.

10.5.1   ANALYSIS OF TIM UNDER NO-LOAD CONDITIONS

Under no-load conditions, the output power delivered is zero, i.e., the output work is zero, although the rotor is rotating, as the output torque is zero. Fig. 10.5 (a) illustrates the minimum current rating of 3.3 A at a working machine at no-load. However, under half-load conditions, as depicted in Fig. 10.5(b), output power delivered to the rotor shaft is half that, compared to rated power. Maximum-rated current of TIM is 4.5 A, with load being adjusted to midway between its minimum and maximum values. Minimum current is 3.3 A in order that half-load current is adjusted to 3.9 A. Fig.10.5 (c) depicts the full-load conditions at which the power delivered to output is maximal. Similarly, Figure 10.5 (d) exhibits blocked-rotor conditions where the rotor of TIM is blocked with a reduced adjustable voltage, fed using an autotransformer. Note that the reduced voltage is of the order of 95 V.

10.5.2   HARMONIC ANALYSIS OF TIM IN ABSENCE OF
HARMONIC ELIMINATION TECHNIQUE

Using data collection and its analysis, harmonics analysis was carried out for TIM using the PHA under different load patterns: no-load, half-load, full-load, and blocked-rotor. Fig. 10.6 (a)-(f) illustrates the harmonic analysis using Fast Fourier Transform (FFT) for line voltages V_1, V_2, and V_3, and line currents I_1, I_2, and I_3 under no-load conditions. It may be noted that 3rd order and 5th order harmonics are 1.1% and 1.1%, respectively, which are more dominant, without using the filter. Overall, THD is 1.8%. Similarly, for line voltage V_2, 3rd order and 5th order harmonics are 1.2% and 0.8%, respectively, and overall THD is 1.7%. For line voltage V_3, 3rd order, and 5th order harmonics are 1.2% and 0.7%, respectively, with overall THD being 1.8%. For line current I_1, 3rd order and 5th order harmonics are 2.7% and 2.5%, respectively, with overall THD being 4.4%. For line current I_2, 3rd order and 5th order harmonics are 5.2% and 3.7%, respectively; overall THD is 7.2%. For line current I_3, 3rd order and 5th order harmonics are 3.5% and 2.0%, respectively, with overall THD being 4.6%.

As shown in Fig. 10.7 under half-load conditions, for line voltage V_1, 3rd order and 5th order harmonics are 1.1% and 1.1%, respectively, with overall THD being 1.9%. For line voltage V_2, 3rd order and 5th order harmonics are 1.4% and 1.1%, respectively. Overall THD is 2.1%. For line voltage V_3, 3rd order and 5th order harmonics

FIGURE 10.5 Voltage-current analysis under (a) no-load (b) half-load (c) full-load (d) blocked-rotor

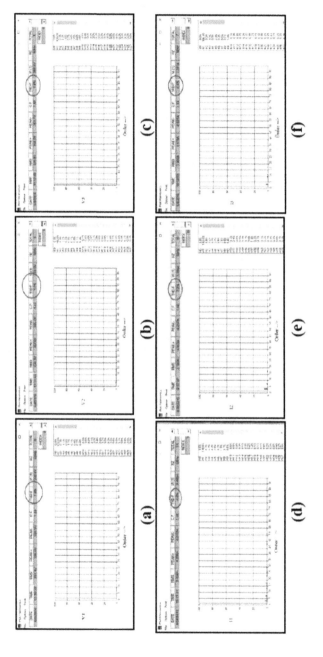

FIGURE 10.6 Harmonic analysis for (a)–(c) V_1, V_2, and V_3 (d)–(f) I_1, I_2, and I_3 at no-load

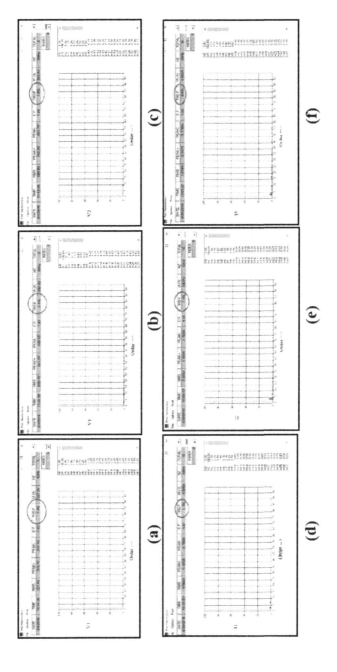

FIGURE 10.7 Harmonic analysis for (a)–(c) V_1, V_2, and V_3 (d)–(f) I_1, I_2, and I_3 at half-load

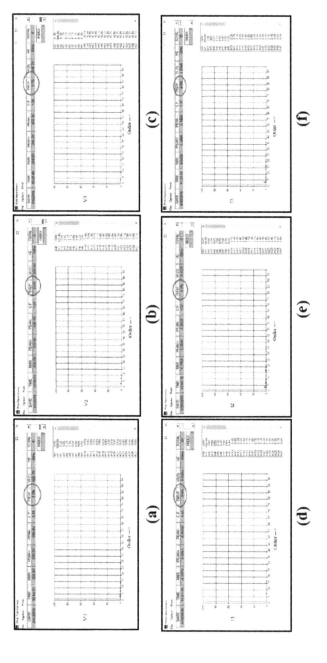

FIGURE 10.8   Harmonic analysis for (a)–(c) V_1, V_2, and V_3 (d)–(f) I_1, I_2, and I_3 at full-load

are 1.2% and 0.7%, respectively; overall THD is 1.8%. Similarly, for line current I_1, 3rd order and 5th order harmonics are 2.5% and 1.5%, respectively, with overall THD being 5.3%. For line current I_2, 3rd order and 5th order harmonics are 5.7% and 3.6%, respectively; overall THD is 7.1%. For line current I_3, 3rd order and 5th order harmonics are 3.7% and 4.0%, respectively, with overall THD being 6.6%.

As shown in Fig. 10.8 under full-load conditions, for line voltage V_1, 3rd order and 5th order harmonics are 1% and 1.7%, respectively, with overall THD being 2.3%. For line voltage V_2, 3rd order and 5th order harmonics are 1.4% and 1.7%, respectively; overall THD is 2.4%. For line voltage V_3, 3rd order and 5th order harmonics are 1.2% and 0.7%, respectively, with overall THD being 1.7%. For line current I_1, 3rd order and 5th order harmonics are 0.5% and 0.3%, respectively; overall THD is 3.9%. For line current I_2, 3rd order and 5th order harmonics are 9% and 2.6%, respectively, with overall THD being 4.8%. For line current I_3, 3rd order and 5th order harmonics are 3.6% and 1.1%, respectively; overall THD is 4.9%.

It must be noted that under blocked-rotor load conditions, the rotor of the TIM is blocked with reduced and adjustable voltage applied, using an autotransformer. As shown in Fig.10.9 under blocked-rotor conditions, the reduced voltage applied is 95 V at 4.5 A maximum current. For line voltage V_1, 3rd order and 5th order harmonics are 90.3% and 1.4%, respectively, with overall THD being 90.9%. For line voltage V_2, 3rd order and 5th order harmonics are 97.4% and 2.7%, respectively; overall THD is 97.9%. For line voltage V_3, 3rd order and 5th order harmonics are 78.3% and 1.6%, respectively. For line current I_1, 3rd order and 5th order harmonics are 0.9% and 0.5%, respectively, with overall THD being 1.4%. For line current I_2, 3rd order and 5th order harmonics are 1.4% and 0.2%, respectively; overall THD is 1.3%. For line current I_3, 3rd order and 5th order harmonics are 1.1% and 0.3%, respectively, with overall THD being 1.4%.

10.5.3 HARMONIC ANALYSIS OF TIM USING PASSIVE FILTER

The passive shunt filter is the most commonly used filter for mitigating current harmonic distortion. In this chapter, the passive filter has been designed to mitigate 3rd order harmonics, 5th order harmonics, and overall current THD. L and C chosen values are 6 mH and 6.5 mF, respectively.

As shown in Fig. 10.10 under no-load test, for line current I_1, it is observed that THD is reduced from 4.4% to 3.3%, using the passive filter. The 3rd order and 5th order harmonics are 2.7% and 2.5%, respectively, and found to be predominant, without using the filter. However, the harmonics are reduced to 1.4% and 2.0%, respectively, by using the passive filter. Similarly, for line current I_2, it is observed that THD is reduced from 7.2% to 6.6%, using the passive filter. The 3rd order and 5th order harmonics are 5.2% and 3.7%, respectively, and are reduced to 4.9% and 3.3%, respectively, using the passive filter. For line current I_3, it is observed that THD decreased from 4.6% to 3.5%, using the passive filter. The 3rd order and 5th order harmonics are 3.5% and 2.0%, respectively and are reduced to 2.5% and 0.9%, respectively, by using the passive filter.

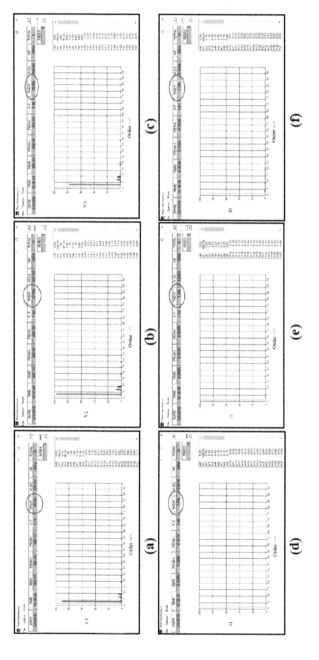

FIGURE 10.9　Harmonic analysis for (a)–(c) V_1, V_2, and V_3 (d)–(f) I_1, I_2, and I_3 under blocked-rotor test

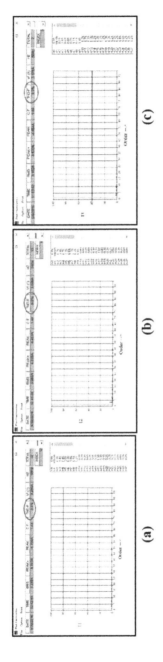

FIGURE 10.10 Harmonic analysis for (a) I_1, (b) I_2, and (c) I_3 using the passive filter at no-load test

For line current I_1 under the half-load test, Fig. 10.11 depicts the harmonics values, using the shunt connected passive filter. It is concluded that THD is reduced from 5.3% to 3.3% using the passive filter. The 3rd order and 5th order harmonics are2.5% and 1.5%,respectively,and found to be predominant, without using the filter, and are reduced to 1% and 1.1%, respectively, by using the passive filter. Similarly, for line current I_2, THD is reduced from 7.1% to 5.6% using the passive filter. The 3rd order and 5th order harmonics are 5.7% and 3.6%, respectively, without using the filter, whereas the values are reduced to 4.2% and 2.7%, respectively, by using the passive filter. For line current I_3, THD is reduced from 6.6% to 4.8% using the passive filter, with the 3rd order and 5th order harmonics being 3.7% and 4.0%, respectively, without using the filter, whereas harmonic level is reduced to 3.4%, using the passive filter.

As shown in Fig. 10.12 under full-load conditions, for line current I_1, it is concluded that THD is reduced from 3.9% to 3.1%, using the passive filter. The 3rd order and 5th order harmonics are 0.5% and 0.3%, respectively without using the filter, but decrease to 0.2% and 0.2% when using the passive filter. For line current I_2, it is concluded that THD is reduced from 4.8% to 4.0%, using the passive filter. The 3rd order and 5th order harmonics are 2.9% and 2.6%, respectively, without using the filter, and are reduced to 2.1% and 2.2%, respectively, by using the passive filter. For line current I_3, THD is reduced from 3.9% to 3.1% using the passive filter. The 3rd order and 5th order harmonics are 0.5% and 0.3%, respectively, without using the filter, but are reduced to 0.2% and 0.2%, respectively, by using the passive filter.

Fig. 10.13 shows the situation under blocked-rotor load conditions. For line current I_1, THD is reduced from 3.9% to 3.1%, using the passive filter. The 3rd order and 5th order harmonics are 0.5% and 0.3%, respectively, without using the filter, but are reduced to 0.2% and 0.2%, respectively, by using the passive filter. For line current I_2, THD is reduced from 1.3% to 1.1%, using the passive filter. The 3rd order and 5th order harmonics are 1.0% and 0.2%, respectively, without using the filter, being reduced to 0.8% and 0.1%, respectively when using the passive filter. For line current I_3, THD is reduced from 1.4% to 0.8%, using the passive filter. The 3rd order and 5th order harmonics are 1.1% and 0.3%, respectively, without using the filter, and decrease to 0.5% and 0.2%, respectively when using the passive filter.

Table 10.1 illustrates THD analysis in current under dynamic conditions of load on TIM in the absence of the passive filter. It reveals that, as the order of load increases on TIM, the level of harmonics decreases and overall THD also decreases. This decrease in harmonics is due to effective implementation of the PWM technique. Table 10.2 illustrates THD analysis in current in the presence of the passive filter. It is clear that the use of the filter has greatly reduced the level of harmonics and overall THD. Table 10.3 shows the comparison of the two modulation techniques in the proposed system. It revealed the overall superiority of DSVPWM over PWM in reducing the level of harmonics in current. DC-offset is the presence of the DC component in AC signals at various PCCs, whereas the level of DC-offset voltage is also found to be reduced, using the DSVPWM technique relative to the PWM technique.

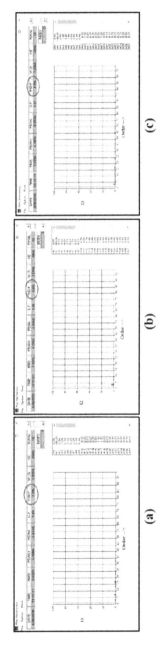

FIGURE 10.11 Harmonic analysis for (a) I_1,(b) I_2, and (c) I_3 using the passive filter at half-load test

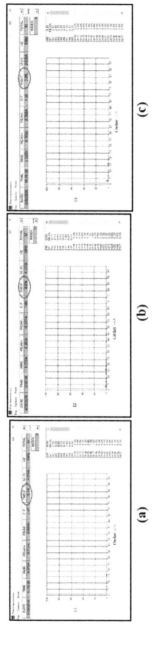

FIGURE 10.12 Harmonic analysis for (a) I_1, (b) I_2, and (c) I_3 using the passive filter at full-load test

FIGURE 10.13   Harmonic analysis for (a) I_1,(b) I_2 and (c) I_3 using the passive filter under blocked-rotor test

TABLE 10.1
Analysis of Current THD (Without Passive Filter)

S. No.	Order of Harmonics	At no-load			At ½-load			At full-load			At blocked-rotor		
		$I_1(A)$	$I_2(A)$	$I_3(A)$	$I_1(A)$	$I_2(A)$	$I_3(A)$	$I_1(A)$	$I_2(A)$	$I_3(A)$	$I_1(A)$	$I_2(A)$	$I_3(A)$
1.	3rd	2.7	5.2	3.5	2.5	5.7	3.7	0.5	2.9	3.6	0.9	1.0	1.1
2.	5th	2.5	3.7	2	1.5	3.6	4.0	0.3	2.6	1.1	0.5	0.2	0.3
3.	I_{THD}	4.4	7.2	4.6	5.3	7.1	6.6	3.9	4.8	4.9	1.4	1.3	1.4

TABLE 10.2
Analysis of Current THD (With Passive Filter)

S. No.	Order of Harmonics	At no-load			At half-load			At full-load			At blocked-rotor		
		$I_1(A)$	$I_2(A)$	$I_3(A)$	$I_1(A)$	$I_2(A)$	$I_3(A)$	$I_1(A)$	$I_2(A)$	$I_3(A)$	$I_1(A)$	$I_2(A)$	$I_3(A)$
1.	3rd	1.4	4.9	2.5	1.0	4.2	3.4	0.2	2.1	2.9	0.7	0.8	0.5
2.	5th	2.0	3.3	0.9	1.1	2.7	2.8	0.2	2.2	0.3	0.2	0.1	0.2
3.	I_{THD}	3.3	6.6	3.5	3.3	5.6	4.8	3.1	4.0	3.8	1.0	1.1	0.8

TABLE 10.3
Analysis of THD and DC-offset, Using SPWM and DSVPWM Techniques

S. No.	Order of Harmonics	Using PWM			Using DSVPWM		
		$I_1(A)$	$I_2(A)$	$I_3(A)$	$I_1(A)$	$I_2(A)$	$I_3(A)$
1.	3rd	1.26	1.08	2.16	0.31	0.45	0.76
2.	5th	0.12	0.07	0.91	0.01	0.05	0.08
3.	I_{THD}	5.27	3.19	5.86	1.44	2.35	2.35
4.	DC-offset voltage	1.426	3.03	4.456	0.8922	0.6975	0.4947

10.6 CONCLUSION

In this chapter, PQ analysis, based on using modulating techniques, is based on ensuring the provision of harmonic-less power to the medical and healthcare sector. The absence of these modulating techniques for controlling TIM drives in healthcare facilities can put the life of patients at risk. The ultimate objective of the proposed systems is to deliver sinusoidal power at constant magnitude. Due to the harmonics present in electric power supply, many complexities involving the harmonics and electro-magnetic interference are produced. Therefore, the proposed systems have attempted to reduce the overall harmonics presented in line current and line voltages at PCC. Harmonic analysis has been carried out for the three-phase SQIM, using the PHA under dynamic loading conditions. It has been revealed that 3rd and 5th harmonics are more dominant. The passive filter has been designed to improve PQ. By implementing the shunt passive filter, 3rd and 5th harmonics have been reduced. Furthermore, PWM and the discrete SVPWM technique are also used to mitigate the current harmonics. Both modulation techniques are coordinated with an AC-DC-AC converter. Observations show that both switching techniques work effectively, although the SVPWM produces better results than the PWM technique. Also, PHA is capable of measuring harmonics up to the 99th order, with a display of 50 harmonics on one screen. There are a number of interesting possible directions for future work based on this research, and these are outlined as follows:

- Harmonic analysis of the induction motor can be carried out using different techniques.
- Harmonic mitigation can be performed by using different techniques.

REFERENCES

1. Kishore A, Prasad RC, and Karan BM, "MATLAB Simulink based DQ modelling and dynamic characteristics of three phase self excited induction generator", *Progress in Electromagnetics Research Symposium*, Cambridge, MA, pp. 312–316, 2006.
2. Ansari AA, and Deshpande DM, "Investigation of Performance of 3 phase Asynchronous Machine under voltage unbalance", *Journal of Theoretical and Applied Information Technology*, 6(1), pp. 21–26, 2009.
3. Bakar MIA, "Assessments for the impact of harmonic current distortion of non-linear load in power system harmonics", *Transmission and Distribution Conference and Exposition: Latin America, IEEE/PES*, Colombia, 2008.
4. Gayathri DB, and Keshavan BK, "A novel hybrid phase shifted-modified synchronous optimal pulse width modulation based 27-level inverter for grid-connected PV system", *Energy*, 178, pp. 309–317, 2019.
5. Chan T, and Shi K, *Applied Intelligent Control of Induction Motor Drives*, Singapore: Wiley, 2011.
6. Sergaki ES, and Moustaizis SD, "Efficiency optimization of a direct torque controlled induction motor used in hybrid electric vehicles", *Proceedings of International Aegean Conference on Electrical Machines and Power Electronics and Electromotion Joint Conference (ACEMP)*, Turkey, pp. 398–403, 2011.

7. Abdelli R, Rekioua D, and Rekioua T, "Performances improvements and torque ripple minimization for VSI fed induction machine with direct control torque", *ISA Transactions*, 50(2), pp. 213–219, 2011.

8. Alsofyani IM, and Idris NRN, "A review on sensorless techniques for sustainable reliablity and efficient variable frequency drives of induction motors", *Renewable and Sustainable Energy Reviews*, 24, pp. 111–121, 2013.

9. Leonhard W, "Controlled AC drives a successful transition from ideas to industrial practice", *Control Engineering Practice*, 4(7), pp. 897–908, 1996.

10. Buja G, and Kazmierkowski M, "Direct torque control of PWM inverter-fed AC motors - A survey", *IEEE Transactions on Industrial Electronics*, 51(4), pp. 744–757, 2004.

11. Chen KY, Hu JS, Tang CH, and Shen TY, "A novel switching strategy for FOC motor drive using multi-dimensional feedback quantization", *Control Engineering Practice*, 20(2), pp. 196–204, 2012.

12. Verrelli C, Tomei P, Lorenzani E, Fornari R, and Immovilli F, "Further results on nonlinear tracking control and parameter estimation for induction motors", *Control Engineering Practice*, 66, pp. 116–125, 2017.

13. Costa BLG, Graciola CL, Angélico BA, Goedtel A, Castoldi MF, and Pereira WCA, "A practical framework for tuning DTC-SVM drive of three-phase induction motors", *Control Engineering Practice*, 88, pp. 119–127, 2019.

14. Ouanjli NE, Motahhir S, Derouich A, Ghzizal AE, Chebabhi A, and Taoussi M, "Improved DTC strategy of doubly fed induction motor using fuzzy logic controller", *Energy Reports*, 5, pp. 271–279, 2019.

15. Ammar A, Bourek A, and Benakcha A, "Nonlinear SVM-DTC for induction motor drive using input–output feedback linearization and high order sliding mode control", *ISA Transactions*, 67, pp. 428–442, 2017.

16. Zemmit A, Messalti S, and Harrag A, "Innovative improved direct torque control of doubly fed induction machine (DFIM) using artificial neural network (ANN-DTC)", *International Journal of Applied Engineering Research*, 11(16), pp. 9099–9105, 2016.

17. Lu S, He Q, and Zhao J, "Bearing fault diagnosis of a permanent magnet synchronous motor via a fast and online order analysis method in an embedded system", *Mechanical Systems and Signal Processing*, 113, pp. 36–49, 2018.

18. Khezzar A, Kaikaa MY, Oumaamar MEK, Boucherma M, and Razik H, "On the use of slot harmonics as a potential indicator of rotor bar breakage in the induction machine", *IEEE Transactions on Industrial Electronics*, 56(11), pp. 4592–4605, 2009.

19. Gyftakis KN, Dionysios VS, Kappatou JC, and Epaminondas DM, "A novel approach for broken bar fault diagnosis in induction motors through torque monitoring in energy conversion", *IEEE Transactions in Energy Conversion*, 28(2), pp. 267–277, 2013.

20. Drif MH, and Cardoso AJM, "Stator fault diagnostics in squirrel cage three-phase induction motor drives using the instantaneous active and reactive power signature analyses", *IEEE Transaction on Industrial Informatics*, 10(2), pp. 1348–1360, 2014.

21. Gunsal I, Stone DA, and Foster MP, "A unique pulse width modulation to reduce leakage current for cascaded H-bridge inverters in PV and battery energy storage applications", *Energy Procedia*, 151, pp. 84–90, 2018.

11 Smart Home (Domotics)

Hemant Kumar Gianey and Sun-Yuan Hsieh

CONTENTS

After studying this chapter, you should be able to understand the concept of the smart home, its emergence and its definition, dimensions, and components in detail. Adoption of the Internet of Things (IoT) in automation is also explained in detail, along with design strategies. In the last section, various issues and challenges, associated with smart homes and different apps, are also discussed in detail. With the assistance of the IoT, we will be looking at the cutting edge innovations that we have encountered or forward to the future ideas that will change our everyday home lives into smart home living. Since its inception, the smart home concept has developed from the execution of explicit tasks to the use of procedures to handle more extensive city challenges worldwide. So, we need to study the concept of the smart home in light of various issues, their challenges and different applications of the smart home idea.

11.1 INTRODUCTION

The IoT involves utilizing various sorts of devices, made by various manufacturers and having different features. From its origins, which focused on individual improvements to the conventional home, exploiting the potential of IoT, the smart home concept has developed from the single building to the development of communities, capable of handling the issues associated with the challenges faced by city living. These days, home automation or robotization frameworks are being utilized to an ever-increasing extent, with many people becoming interested in making their homes more user friendly and secure. On one hand, these frameworks provide increased security and peace of mind; on the other hand, these frameworks, when presented as business structures, improve not only security and comfort, but also permit centralized management. Therefore, these systems contribute to overall reductions in cost and increases in energy and resource efficiency, which are really important issues nowadays.

Some current smart home frameworks depend on wired innovation; a standard automation or robotization framework does not take long to install if the framework has been planned prior to the build and inserted throughout the development of the building. If an existing building needs to have the automation system retrofitted, this needs a lot of effort and cost, since cabling is critical and costly. Wireless systems could assist here. In recent years, different wireless advances have occurred. Frameworks dependent on wireless advances are being utilized each day and everywhere. A comparative review of wireless norms has been carried out, and such information will provide significant help to anybody looking for the most appropriate framework for given tasks.

11.1.1 Concept of Smart Home

A smart or sensible home allows users to achieve increased resource efficiency and make financial savings by managing lights, window covers, and by monitoring. Various technologies from the smart phone have extended the users' interest in management of major devices. The machine-manageable device allows users to perform tasks before reaching the home/office. A smart home system provides solutions for particular tasks using various technologies, particularly valuable for the physically disabled or older, less able people, using smart phone apps. According to a report, almost 70% respondents said that self-adjusting thermostats and doors that could be locked remotely from a distant location are the most important functions among the smart home gadgets.

The key advantage, resulting in the home-owner inclining toward this system over equivalent non-smart sorts of devices, is that the alerts sent by the smaller-scale controller associated with Wi-Fi are regularly received by the user or customer on the phone from any distance, whereas long-range interpersonal communication sites like Twitter can additionally be exploited to send message alarms to the client.

A smart home system manages the integrated systems in the home. The modules in these systems are managed by users using interface devices like a smart phone. The project emphasizes essential aspects of our everyday lives. Various sensors could be managed and monitored and can be adjusted easily. The system can be accessed locally or remotely from any versatile or hand-held device associated with the web. Wi-Fi development is selected to be the framework establishment that interfaces the server with the sensors. Wi-Fi is selected to update the framework security and to extend system flexibility and versatility. The home mechanization framework has the ability to deal with the following sectors in the client's home and to send alerts: temperature, motion detection, fire and smoke identification, lights, fans, air conditioner system, water-pump, and so forth. The framework has a straightforward plan and program, and it requires hardly any input from the client, yet it is customized for the client's individual needs.

11.1.2 Needs of a Smart Home

The primary objective of a smart home system, using IoT, is one which is capable of managing household appliances through an easily handled interface, providing an automated system to manage home devices wirelessly, effectively, and efficiently, with the following goals:

  i. To deliver an uninterrupted output.
  ii. To manage home devices *via* a button.
  iii. To manage the system through Wi-Fi.
  iv. To offer an extensible platform for future expansion and enhancement.
  v. To optimize a critical monitoring system.

We are living in a time with so many technological advancements, which make our lives simpler; this is a blessing, considering the fast-paced lives most of us lead. In short, the technology allows the home user to manage his/her home even when away *via* a smart-phone or an iPad. A number of reasons for adopting a smart home are listed below, with ways they could benefit us (Figure 11.1):

  i. Safety and peace of mind:
  ii. Security and access management
  iii. Energy efficiency
  iv. Comfort and well-being
  v. Convenience and efficiency

11.1.3 Smart Home Outcomes and Deliverables

The following are the outcomes of the smart home system, which are designed based on the requirements and specifications of the client:

  i. **Flexibility for new appliances and devices**: Smart frameworks will, in general, be versatile with regard to the accommodation of new devices and technological advances. It does not make a difference with respect to how dynamic your devices are these days, as there will be more up-to-date, progressively noteworthy models created in the future.
  ii. **Maximizing home security**: When fusing security and surveillance is included in the smart home system, security should soar. There are huge numbers of alternatives possible, several of which are now being investigated.

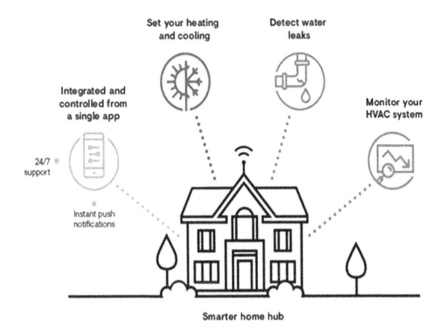

Set your heating
and cooling

Detect water
leaks

Integrated and
controlled from
a single app

Monitor your
HVAC system

24/7
support

Instant push
notifications

Smarter home hub

FIGURE 11.1   A smart home

iii. **Remote oversight**: Managing all at-home devices from one location, the comfort issue is enormous. Having the alternative to keep the sum of the advances in your home controlled by one interface could be a gigantic step forward for development and domestic organization.

iv. **Extend effectiveness**: Depending on how we design a smart home, it is possible to make it able to face future threats and development better.

11.2   THE EMERGENCE AND DEFINITION OF THE SMART HOME

In this section, the smart home framework, the apparatuses, the various capacities, prerequisites, and different variables, are discussed to give a thorough portrayal of the smart home framework.

11.2.1   EXISTING SYSTEM AND SOLUTIONS

The current IR or Bluetooth remote oversights present in the market are, by and large, device specific, and the equivalent devices couldn't be utilized properly. The frameworks dependent on Global System for Mobile (GSM) are expensive and similarly moderate in terms of the devices which can be controlled. Electrical devices which are controlled through Bluetooth cannot be supervised from a distant location. The problem with current home security/observation systems is also associated with the fact that clients are frequently away from home for long periods (Table 11.1).

TABLE 11.1

Literature Survey and Findings

Paper Title	Tools/Technology	Findings	Citation
Smart home research for the elderly and the disabled	AI, Machine Learning	Studies human neurology via machine learning and acts via AI	Vancouver, BC, Canada
Agriculture monitoring system	Arduino	Temperature and soil moisture monitoring process, using Android-based smart phone	N. M. Z. Hashim, S. R. Mazlan Centre for Telecommunication Research & Innovation (CeTRI)
Smart home activities	Electric power components and systems	Electric power components and system, using Android-based smart phone	University of Waterloo, Waterloo, Canada
Control home appliances from the internet, using Arduino and Wi-Fi	Arduino, Wireless	Control of home apparatuses from anywhere in the world, via the internet.	Rajeev Ranjan, Shambhoo Kumar
Ageing and technology: a review of the research literature	AT and ICT intervent ions Z-wave, Zig-Bee	These frameworks offer individuals, who are less mobile or in poor health, the chance to be more independent, instead of remaining in a supported-living room.	Christina M. Blaschke, Erin E. Mullen, The British Journal of Social Work, Volume 39
Smart home automation security	Bluetooth, GSM, GPRS, RFID	Security flaws in existing home automation systems.	Arun Cyril Jose, Reza Malekian, University of Pretoria, South Africa
Analytical framework for research on smart homes and their users	IoT, ACT	The expository system, connections between explored subjects and customs	T. Hargreaves and C. Wilson, Smart Homes and Their Users Human–Computer Interaction Series
Mobile-based home automation using the Internet of Things (IoT)	Bluetooth, IoT, Mobile computing	Smart home computerization idea, utilizing minimal effort of the Arduino board	Kumar Mandula, Ramu Parupalli, CH.A.S. Murty, E. Magesh. Centre for Development of Advanced Computing
Communication and control for a home automation system	Arduino, AI	This project controls with the help of the sound of the robot	C. Douligeris; J. Khawand; C. Khawand Dept. of Electr. & Comput. Eng., Miami Univ., Coral Gables, FL, USA
Greenhouse automation	ZIGBEE	Optimization of environmental conditions for better plant growth	Y.R. Dhumal, J.S Chitode Electronics Department and Bharati Vidyapeeth University.
Smart home research	AI	Maximum implementation on a low budget	Li Jiang; Da-You Liu; Bo Yang, Shanghai, China

11.2.2 IDENTIFIED PROBLEMS

A major issue recognized in past investigations into the existing smart home framework is that, in the GSM-based framework, in the event that General Packet Radio Services (GPRS) is not accessible, the system will not work. Most of the current smart home systems use the GPRS structure, which is expensive. Some frameworks use Wi-Fi connections, yet, for the most part, they use Raspberry Pi for computing, the cost of which is also exorbitant. All things considered, the home computerization or robotization system does not offer some obvious options like management of outside lights.

11.2.3 TOOLS AND TECHNOLOGIES USED

There are various strategies to regulate home devices, such as IoT-based home mechanization/robotization over the cloud, home computerization under Wi-Fi through the use of any cell phone, Arduino-based home automation and robotization, using progressed supervision, and radio frequency (RF)-based home robotization structures (Table 11.2).

There are different Wi-Fi innovations like Wi-Fi 802.11a, 802.11b, 802.11g and 802.11n. Compared to other wireless communication techniques, NodeMCU possesses the benefits listed below:

a) Low energy consumption;
b) Low cost;
c) Flexible
d) Fast, easy deployment
e) Security
f) Product interoperability;

11.3 DESIGN AND COMPONENTS OF SMART HOME SYSTEM

The smart home system consists of micro-controllers and sensors connected by way of the internet, to access home devices remotely *via* an application. Message alerts

TABLE 11.2
Consolidated Comparison Reports of all Systems

System	Communication	Access	Number of Devices	Speed	Real time
GSM	SMS	Anywhere	Unlimited	Slow	No
Bluetooth	Bluetooth instructions	Restricted	Unlimited	Fast	Yes
Phone based	Phone lines	Anywhere	12	Fast	No
Wireless	Radio, IR	Contingent upon range and spectrum of waves	Unlimited	Fast	Yes

(regarding the current status of modules) will be prompted and also the user will get updates through Twitter. The user could give manual instructions as well as voice instructions to manage home devices.

The framework is made up of one master or ace module and numerous slave modules. A test run has been carried out with one smaller-scale ace controller, and miniaturized slave controllers. The miniaturized ace controller makes it conceivable to deal with each module in the family unit, whereas the smaller-scale slave controllers adhere to directions given from the ace. The ace module has route catches; the ace module is introduced on the client's process and data automation (PDA), from which the client could deal with all the other slave modules, the number and dispersion of which are dependent entirely upon the client. Slave modules are recognized by their 'slave ID' number. The individual slave modules are indistinguishable, in terms of both the equipment and the programming. This strategy is exceptionally basic and makes manual 'reconstructing' of the slave modules conceivable, without alteration to the module's product (Figure 11.2).

The user needs to first register instructions to take control of the devices. After signing in, the user could choose to either update or simply view the current status of a particular device. Each device is assigned some trigger values for turning them on/off or to manage them. The user could also add further devices and manage them by an app or through voice instructions. A notification message or update is sent to the user *via* Twitter. The smart home system also provides a method to update the social media site.

11.3.1 ENVIRONMENTAL, ECONOMIC AND SOCIETAL BENEFITS

There are several benefits, which are listed below and illustrated in Figure 11.3:

i. **Saving Energy and Increasing Comfort**: There are numerous motivations to utilize a smart home framework; however, probably the best explanation is that it helps to save energy. A home that saves energy is not just going to help the client save money but it also contributes toward securing the Earth by utilizing less of the world's limited resources. It might be difficult to accept, yet utilizing less energy could really improve your peace of mind. Using a smart home framework, you could oversee everything remotely, which means that, when you're not home, you could switch off any device by means of your PDA. Also, you could simply switch them back on when you need them.

ii. **Heating and Cooling**: One of the best aspects of owning a smart home framework is having the option to oversee and program your indoor environment remotely. An appropriate home framework system could take in environmental data and manage the temperature, etc. as necessary. Associated indoor regulators will utilize data on parameters like relative humidity, outdoor air temperature, whether the house is occupied or not, energy cost, and dampness to give you the most agreeable indoor temperature.

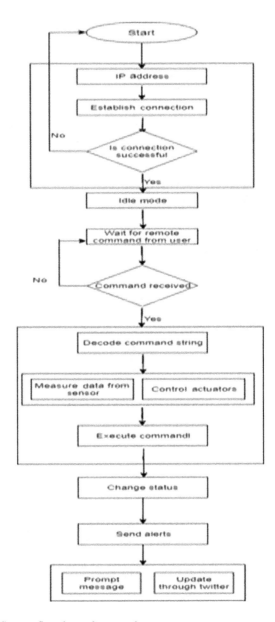

FIGURE 11.2   System flowchart of a smart home

iii. **Lights**: Ever go out and realize in the car on the way to work that you had left the lights on? In the event that you have a smart home framework, you could switch off the lights before you even finish your journey. Your advanced mobile phone could deal with any light in the house, allowing you to save energy that you would prefer not to waste.

FIGURE 11.3 Benefits of smart homes

iv. **Devices and Electronics**: Many of your home's devices don't have to run throughout the day, which would expend a lot of energy. Such energy wastage costs the client money and damages nature and the environment. A smart home framework gives you oversight over all the hardware and electrical devices. When you aren't using these things, there's no purpose in them being on, consuming energy. You may be astonished by how much energy you utilize each day, and how much your energy use can be reduced by including and using a smart home framework.

Climate change is a growing problem that requires immediate and aggressive solution. Using smart home automation technology, millions of homes around the world could contribute to bringing about positive environmental change. Many smart home devices could also help consumers make well-informed decisions when buying new appliances.

11.4 SMART HOME ARCHITECTURE

The smart home mechanization/robotization system involves three essential modules: the server, the hardware interface module, and the software package. The customer may use an appropriate device with which to log-in to the server application. If the server is connected to the web, remote users or customers could experience a rapid response (Figures 11.4 and 11.5).

At the point when the client opens the graphical user interface (GUI) app, he/she sees a primary page alongside a list of devices associated with the organizer box, and barrier insights, regarding their present status. From this list, the client can choose the device he/she needs to oversee. After a device is chosen, the client can see all the highlights associated with that device, that could be modified through that app.

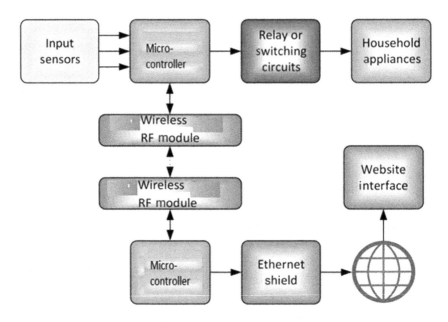

FIGURE 11.4   Block diagram of a smart home management system

In the wake of altering particular highlights, the client could utilize the "Modify" button to send a report about these changed highlights to the home devices. In addition to making component changes in the devices, the client could likewise utilize "Turn Off" or "Turn On" buttons directly with the device itself or to start-stop its operation over time. As a controller, we are utilizing a peripheral interface controller (PIC)-fueled improvement board to deal with the devices. On this oversight board, the clients can modify the devices as indicated by their needs. These changes should also be carried out, using the app, through the internet. The data are then sent to the framework, which responds to secure the synchronization inside the framework and the controller.

11.5   ISSUES AND CHALLENGES OF A SMART HOME

In researching this literature review, it became apparent that the majority of the literature on smart homes centered around security issues. Discussions of a large number of security issues have been rehashed by various authors over the years. We could find no paper that completely covers the entire architecture of a smart home. In addition, the main issue was to have a low-cost system which was affordable, easy to install, easy to operate, and which could be extended. Home automation systems faces challenges such as the significant expense of-usership and security.

A progressive electronic home automation/robotization framework could be achieved in the future to guarantee high levels of security, which could be overseen, *via* the web, by means of high-definition HD spy cameras. Alongside this, the framework could be set up in any domestic or work-related building, and could be viewed

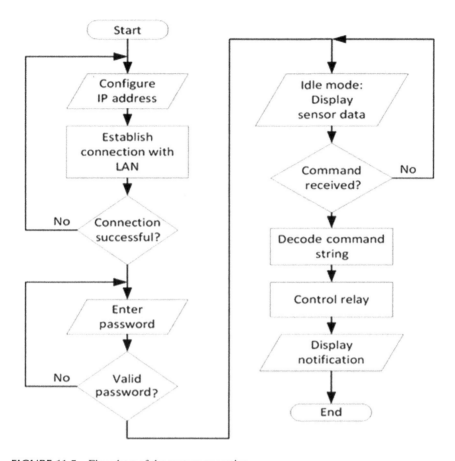

FIGURE 11.5 Flowchart of the system operation

from any place. The benefits of smart building robotization, listed below, could be particularly useful:

i. The different future applications might be utilized to oversee different family devices *via* the internet
ii. Industrial automation and management.
iii. Enhancement of security to cover issues in extremely restricted areas.
iv. Machine-driven fire exit system.
v. Thermostats to manage temperature, change mode, change fan mode to on/auto/circulate, read, and adjust humidity (Figure 11.6).

The following are the assumptions made for the smart home system:

a) The client will interface with an app, will enter settings to oversee or change the status of devices and to check on their current status.

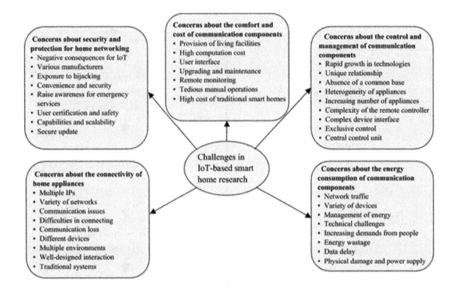

FIGURE 11.6   Issues and challenges of smart homes

b) If the battery is low, at that point, a warning message will be sent to the client.

c) Sensors are exceptionally dependable and work appropriately.

d) Moisture sensors sense the dampness level of the soil.

e) The system cannot be changed by unapproved persons.

f) Security should be ensured, with communication among each of the sensors and modules being carried out easily, with no obstruction.

g) Only the client's voice to be recognized and authorized to make changes.

11.6   CONCLUSION

The smart home framework is a secure, universally open, auto-configurable, remotely overseen arrangement requiring minimal effort from the client. The methodology described in this chapter section is novel but has been designed to accomplish the objectives to oversee home devices, by remote use of Wi-Fi, fulfilling the client's needs. Wi-Fi allows the system to be overseen or modified remotely, provides home security and is "smart", in comparison with existing home security measures. Subsequently, we can infer that the necessary objectives and goals of the home computerization framework have been accomplished. The framework plan and design have been examined in this chapter, and the model introduces the fundamental degrees of home machine oversight and remote supervision. Finally, the proposed system is better than the commercially available home robotization structures from the points of view of flexibility and versatility.

11.7 FUTURE WORK

Progressively, the electronic home automation/robotization framework could be expanded in the future to guarantee high security; to carry out smart waste management; distributed storage could be utilized; various social media websites could be utilized to send warnings or to receive messages. Alongside this, the framework could be consolidated in an entire structure of any organization or private structure, and could be viewed remotely from any place. Along these lines, the most favorable circumstances of home computerization/robotization could be achieved, to benefit the client even more.

BIBLIOGRAPHY

1. A. Amudha. 2017. Home Automation using IoT. *International Journal of Electronics Engineering Research*, 9(6), 939–944.
2. B. Ciubotaru-Petrescu, Chiciudean, D., Cioarga, R., Stanescu, D. 2006. Wireless Solutions for Telemetry in Civil Equipment and Infrastructure Monitoring. In: *3rd Romanian-Hungarian Joint Symposium On Applied Computational Intelligence* (SA CI), May 25–26.
3. Anamul Haque, Kamruzzaman, Sikder, Ashraful Islam, Md. 2010. A System for Smart Home Manage of Gadgets Based on Timer and Speech Interaction.
4. N. P. Jawarkar, Ahmed, V., Ladhake, S. A., Thakare, R. D. 2008. Micro-Controller Based Remote Monitoring Using Mobile Through Spoken Instructions. World. *Journal of Management Science and Engineering*, 2(1), 6–11.
5. Kok Kiong Tan, Lee, Tong Heng, Soh, Chai Yee 2002. Internet-Based Monitoring of Distributed Manage Systems-an Undergraduate Experiment. *IEEE Transactions on Education*, 45(2).
6. Matthias Gauger, Minder, Daniel, Wair Conditionerker, Arno, Lair, Andreas 2008. Prototyping sensor-actuators networks for home automation. Realwsn', Glasgow, United Kingdom.
7. Nikhil Singh, Bharti, Shambhu Shankar, Singh, Rupal, Kumar, Dushyant 2014. Remotely Managed Home Automation System. In: *IEEE International Conference On Advances In Engineering & Technology Analysis, 2014*, Dr. Virendra Swarup Group of Institutions, Unnao, India.
8. I. Potamitis, Georgila, K., Fakotakis, N., Kokkinakis, G. 2003. An Integrated System for Smart-Home Manage of Gadgets Based on Remote Speech Interaction Eurospeech 2003. In: *8th European Conference On Speech Communication And Technology*, pp. 2197–2200, Geneva, Switzerland.
9. Rozita Teymourzadeh, Ahmed, Salah Addin, Chan, Kok Wai, Hoong, Mok Vee. 2013. Smart Gsm Based Home Automation System. In: *Conference On Systems Process And Management (ICSPC)*, December 13–15, Kuala Lumpur, Malaysia.
10. Usb Wi-Fi (802.11b/g/n) Module Along With Antenna For Raspberry Pi, *Adafruit*, 2014. https://Www.Adafruit.Com/Products/1030.
11. Xbee Pro Module - Series 1 - 60mw Along With Wire Antenna - Xbp24-Awi-001, *Adafruit*, 2014. https://Www.Adafruit.Com/Products/964 (2014).
12. Rishabh Hemant Kumar Gianey Rishabh. 2019 August. Implementation of Internet of Things and Protocol. In: *Springer International conference on Inventive Computation Technologies*, Coimbatore, 272–278. doi:10.1007/978-3-030-33846-6_31, 2020.

13. Hemant Kumar Gianey, Adhikari, Mainak. 2019 April. Energy Efficient Offloading Strategy in Fog-Cloud Environment for IoT Applications. In: *Internet of Things: Engineering Cyber Physical Human Systems*, Elsevier. doi.org/10.1016/j. iot.2019.100053.

14. Rajeev Ranjan, Shambhoo Kumar. 2017. Control Home Appliance from Internet Using Arduino and WiFi, November 17. https://www.hackster.io/iotboys/control-home-appli ance-from-internet-using-arduino-and-wifi-f65e10#overview.

12 Developing Ecosystem for Tracking Vehicles in Disaster Prone Areas

Vivek Kaundal, Raghav Ankur, Tracy Austina Zacreas, Vinay Chowdary, and Asmita Singh Bisen

CONTENTS

12.1 INTRODUCTION

Faster rescue and evacuation operations in post-disaster regions are very challenging tasks, due to the uncertainty associated with localizing human beings. Additionally, failure of the pre-existing communication networks in that region prevents the trapped individuals from providing information on their location to rescuing authorities. This leads to further delays in rescue operations. The famous Brundtland Report [1] on sustainable development strongly emphasizes action on several key issues, amongst which life preservation in the face of natural disasters is one of the prominent issues.

Much importance is also given to the existing cellular network to make it stronger during such post-disaster situations, where the emphasis will be on a gradual shift from macrocell strategy in the traditional cellular network to heterogeneous topology [2]. However, this is driven by a need to increase channel capacity. With the increase in channel capacity, both data transmission speeds and number of users increases. But this tells us little about the confidence with which communication can take place under specific circumstances. Hence, under total link failure conditions, there will not be any communication, despite the capability of high data speeds and the capacity for many users per channel. The point is not to challenge the modern cellular topologies, but to emphasize the capability of heterogeneous cellular networks, specially designed for their use in disaster management. Under such circumstances, the link reliability or localization accuracy is of utmost importance, rather than data speed.

In the research investigation described in [3], emphasis was placed upon the increase in human quality-of-life for sustainable development, by using wireless sensor networks (WSNs). The work presented in [4] highlights the significance of post-disaster action planning in the context of sustainable development, where it is explained that the locals in the area should participate and ask for supply of their immediate needs and participate in decision making. However, the research work described in [5] presents the distinct types of planning schemes in the post-disaster scenario, emphasizing sustainable hazard removal rather than recovery. Likewise, the increasing need for wireless sensor networks for improved resilience post-disaster has been emphasized [6]. The author in [7] presents a detailed account of remote-sensing applications in disaster management, whereas another study focused on the need for adaptation of new methods for disaster management, including the redundancy of the rescue systems and increased communication amongst managers [8]. In [9], the authors highlighted the need to develop communication systems for disaster management. The simulation results depict that WSN will play a significant role in sustainable development where the emphasis is put on managing the post-disaster conditions. Therefore, the adoption of new methods can increase the safety factor of humans trapped in post-disaster condition.

Wireless technology plays a prominent role in disaster management and many devices have been designed that provide a warning before the disaster, thus saving the lives of individuals in disaster-prone areas [10–12]. WSN is a very promising communication network system [13] in the field, especially where a standalone network has to be implemented in the disaster-prone area to get a pre-indication of the disaster. Also, WSN has been extensively applied in the field of environmental monitoring [14], *ad-hoc* vehicular networks [15], and localization of targets [16].

Localization is one of the important applications of WSN, which is of particular interest in this research area. Some of the methods for localization are listed as Global Positioning System (GPS), Angle of Arrival (AOA), Time of Arrival (TOA), Time Difference of Arrival (TDOA), and Received Signal Strength Indicator (RSSI). All of these techniques come under the category of range-based localization. RSSI-based localization is very promising in terms of the cost involved in designing the system. In this chapter, a customized mobile sensor node has been designed that focuses on RSSI-based localization. The ZigBee S2 series, integrated in the mobile sensor node, has a unique feature of sending a RSSI signal, whenever requested. Also, the cost involved for designing such nodes is minimal and, in terms of power consumption and reliability, ZigBee-based mobile sensor nodes show the best results in achieving localization. Post-disaster scenarios involve further risk to lives due to the absence of a proper responding system between those who are trapped for communication between them and the rescuing authorities, as every communication link is based on the infrastructure which is now non-operable. For localization, GPS is a mature outdoor communication technology but works inefficiently in indoor environments and has noisy access in mountainous areas. Moreover, inclusion of hardware of GPS which are cost ineffective. Unmanned aerial vehicles (UAVs) are used for real-time searching for and monitoring of individuals in disaster-prone areas [17], but they are lacking in terms of power consumption (only short flight times are possible) and accessibility at night [18,19]. Furthermore, flight time is not enough to cover more than a limited geographical region.

The demand for WSN is centered on the capability of the node for data transmission with optimized routing algorithms. Apart from the above-stated way of localization, there is also a need to optimize the localization techniques. Therefore, a literature survey was conducted for the optimization techniques for localization [20]. The authors implemented the Particle Swarm Optimization (PSO) algorithm for the efficient localization of the nodes [20]. The fitness function was designed by using the trilateration technique, and 100 particles were used to localize a node. Furthermore, RSSI has been used as a tool for the localization of nodes and a back-propagation algorithm is implemented to optimize the objective function [21].

The MATLAB simulation environment has been used for a 100 m × 100 m area. In another study, RSSI was implemented to reduce the noise interference of Link Quality Indicator (LQI), where the detected object is trained by the radial basis function network [22]. It uses LQI for localization for an indoor study area of 7.26 m × 16.5 m. The authors put greater emphasis on sensor node position estimation and optimization methods to reduce the error from multipath propagation [22]. Another approach, based on wavelet-based features (WBF), has been proposed [23] for a mobile station to be localized in an internal or indoor environment facility. However, the authors considered only indoor environments and focused mainly on finding distance only. In [24], a ubiquitous method is proposed to localize the WSN nodes, being compared with classical approaches to RSSI-based localization.

The categorization of the literature review conducted for this chapter is shown in Fig. 12.1, to provide a broader perspective.

So, based on the survey, below are the points covered in this chapter:

   i. Existing technologies for localization.
  ii. Proposed indigenous sensor network ecosystem
 iii. Proposed technique and algorithm
     a) VPM (Vector Parameter-based Mapping) Protocol
     b) Hybrid TLBO (Teaching–Learning-based optimization) and unilateral technique

Classification of literature survey

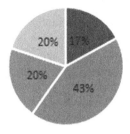

- Literature on disaster management
- Literature on node designing and protocol for disater management
- Literature on algorithm designing
- Literature on sustainable development

FIGURE 12.1   Categorization of the literature review

   iv. Experimental test bed set-up in a location, and design of sensor nodes
   v. Validation and results of the modified approaches for localization
  vi. Conclusion and future scope.

12.2   EXISTING TECHNOLOGIES FOR LOCALIZATION

The trilateration algorithmic is a best-known traditional approach for localization of a node (Fig. 12.2). In the conventional trilateration algorithm, as shown in Fig. 12.2, there are three nodes, r1, r2 and r3, in fixed locations forming a triangle, with the middle node being the trapped node; using the Euclidian distance equation, the location of the trapped node can be estimated easily. Similarly, there is another approach to localization, namely location fingerprinting (Fig. 12.3). The location fingerprinting algorithm captures all RSSI values and represents those RSSI values on a logarithmic scale. The captured RSSI values are called a radio map and the array of the same is known as Radio Frequency (RF) signatures or fingerprinting. So, for the first location L1, the array is represented as

$$ssL_1 = [ssL_{l1} \; ssL_{l2} \; ssL_{l3} \ldots\ldots\ldots ssL_{li}]$$

In this chapter, the sensor network will be designed in a well-known predefined disaster-prone path with the help of the technique of location fingerprinting. The RF signatures of all the vehicles moving all the way through the disaster-prone area are captured and will be communicated to the main central node. So, when the disaster happens, any vehicle can easily be tracked, using the appropriate RF signature.

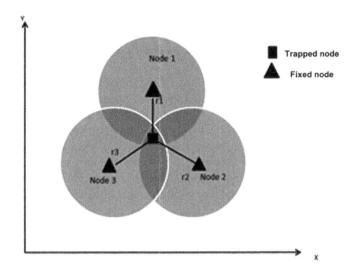

FIGURE 12.2 Trilateration Algorithm

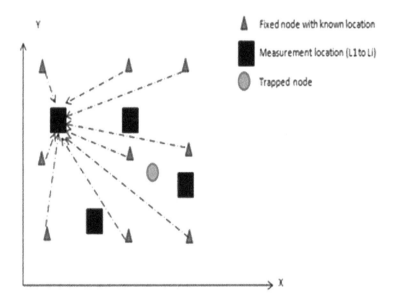

FIGURE 12.3 Location Fingerprinting Algorithm

12.3 PROPOSED INDIGENOUS SENSOR NETWORK ECOSYSTEM

Firstly, a network must be established in the disaster-prone area, using a fixed node, known as the anchor node, a visitor node (movable node) and a rescue operation node. The fixed node will be deployed in the predefined path and acts as a router node, which will direct the information from the visitor node location to the main central node. Visitor nodes are associated with visitors and act as coordinator nodes, which will continuously send the data to the anchor node (fixed node); the data refer to the RSSI. If the disaster happens, then the third node, known as the rescue operation node, will start the rescue process, acting as an end-device node. The fixed nodes are sufficiently intelligent to obtain the information from vehicles passing across it and making a radio map. As the vehicles are already registered at the registration point (see Fig. 12.4) far away from the disaster-prone area, updated information regarding the estimated location of vehicles, and also the number of vehicles, can be obtained at each and every point. Fig. 12.4 shows the block diagrams of the system and will show the roadmap of the disaster-prone area where the network is going to be deployed. The same information is routed through the fixed nodes to the registration point. So, a radio map can be designed at the registration point. As a result, in situations where disaster strikes, and fixed nodes become damaged, then the updated estimated location is safe at the registration points. The proposed system starts from the authentication point in the disaster-prone area i.e. Registration point 1 (Fig. 12.4), where the vehicle entry gets registered and the vehicle gets the wireless node. During the registration process, the number of persons and the vehicle number that is going to travel through that disaster-prone area will be noted down.

After the registration process in the wireless node, the vehicle enters the disaster-prone area and the wireless network is initialized. There are anchor nodes already present, which are communicating with the node present in the vehicle. The estimated location of the vehicle is updated all the time and data are logged wirelessly to the Registration point. As soon as the vehicle crosses the disaster-prone area endpoint, the smart device in the vehicle informs Registration point 2, and the data

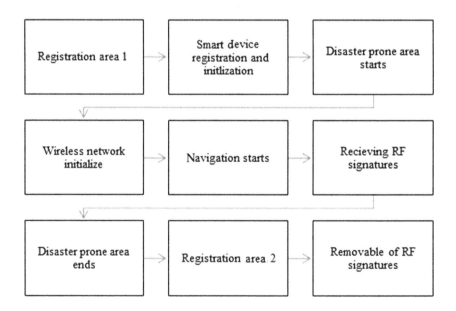

FIGURE 12.4   Basic setup of proposed Architecture

will be removed. This technique can make a radio map of the vehicle node. If a disaster happens, then the vehicle node will become the trapped node. The trapped node is now automatically localized with the rescue team node. The last-estimated location can be tracked by radio map and, after obtaining the estimated location, further localization will be done by the unilateral technique, using VPM (Vector Parameter-based Mapping) [25]. The overall setup of location fingerprinting can be better understood by reference to Fig. 12.5.

12.3.1 PROPOSED TECHNIQUE AND ALGORITHM

The aim of the algorithm is to reach the trapped node by the VPM (Vector Parameter-based Mapping) protocol. Firstly, the node will scout for the RSSI and thereby calculate the estimated distance. The RSSI results are stored in the local memory of the rescue team node. The node will calculate the least remote RSSI value and the information to search for the new location, to move toward the trapped node. The rescue team node continues its steps until it reaches the trapped node. The algorithm is independent of the other anchor nodes, unlike in the trilateration algorithm. In this section, we will first understand the VPM protocol and then the hybrid algorithm to achieve optimization of the designed algorithm.

12.3.2 VPM (VECTOR PARAMETER-BASED MAPPING) PROTOCOL

The VPM protocol is state-of-the-art design, based on the unilateral technique. The blue dot is the trapped node and the red dot is the node with the rescue or searching team (refer to Fig. 12.6). The disaster-prone area, where the node is trapped, is

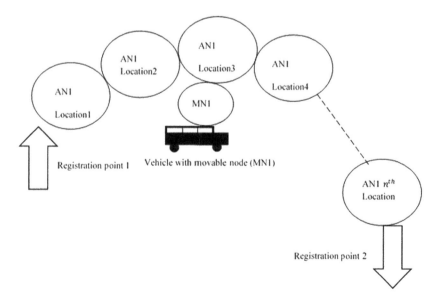

FIGURE 12.5   Location fingerprinting

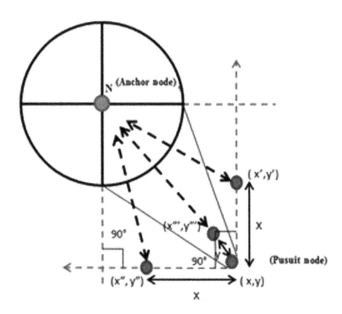

FIGURE 12.6   Approach toward Anchor Node-VPM Protocol-based Mapping

divided into the vectors. Let us surmise that the current location of the searching node is in position (x, y). Now the best possible move of the node is 90° in a direction toward (x', y') or (x'', y''). There is no situation where it can move to an angle greater than 90°, because this would lead to an increase in RSSI value. So, the maximal angle of movement will be 90°. If the searching node moves toward (x', y') or (x'', y'') to a distance x or moves toward (x''', y''') to a distance y, then it will obtain the same RSSI value, due to the omnidirectional behavior of the Xbee antenna. An important point to note is the distance $y \prec x$, so the node gets a stronger RSSI signal at (x''', y''') than at the points (x', y') or (x'', y''). So, the most tenable move of the searching node is toward (x''', y'''), i.e., to an angle of 45°. Practically it will always be α° <90°. So, the new coordinates of the searching node will be (x''', y'''). Again, the searching node maps the area through vectorization and moves accordingly toward the trapped node by an α° angle which is always less than (<) 90°. The rescue team node (pursuit node) always does the search training in +x, −x, +y and −y directions, and moves accordingly toward the anchor (target) node (refer to Fig. 12.7).

The rescue team node (pursuit node) always does the search training in +x, −x, +y and −y directions, and moves accordingly toward the anchor (target) node (refer to Fig. 12.7).

12.3.3   HYBRID TLBO (TEACHING–LEARNING-BASED OPTIMIZATION) AND UNILATERAL TECHNIQUE

Localization of the node is carried out using two methods, i.e., LNSM (Log Normal Shadowing Method) and Hybrid TLBO-unilateral algorithm [25]. This hybrid

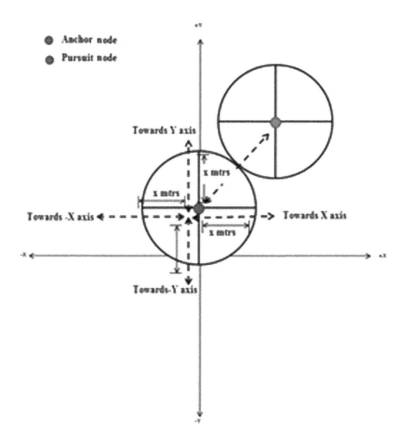

FIGURE 12.7   Training pattern in –X, X, –Y, Y direction, using VPM Protocol

algorithm is based on the classical approach of the TLBO algorithm [26] and a modified trilateration technique, i.e., unilateration, using the VPM protocol, which has already been explained in the previous section (Section 12.3.2).

In this section, RSSI values have been used for the localization process, as discussed in the literature survey. It has been observed during the experiments that the RSSI values fluctuate considerably or sometimes they fade too. To remove the problem of noisy channels, it is very important to model the wireless channel. For the LNSM approach, two nodes have been placed in outdoor locations. One of the nodes acted as an anchor (fixed) node and the other one as a pursuit node. Both nodes have been equipped with Xbee transceiver modules and are able to obtain the RSSI values. The RSSIs are captured in SD card module (in-built features of self-designed nodes). The pursuit node is put through to move around the anchor node and the estimated distance is calculated using LNSM, by obtaining the RSSI values. Using LNSM, the distance can easily be calculated, and localization of the nodes can be achieved. Once the nodes are modeled, then the localization technique is optimized further, using the hybrid TLBO-unilateral algorithm.

First, let us understand the classical approach of the TLBO. TLBO is a teacher–learner-based algorithm that mainly focuses on the learning given by the teacher node to the learner node, so that the best learner node will become the teacher node for other nodes. This will be explained further in detail later in this chapter and the unilateral algorithm is the improved version of the trilateration algorithm. So, in both experiments, RSSI values have been obtained and, through the RSSI values, the pursuit node is subject to discovering the trapped node.

12.3.3.1 Modeling of Wireless Channel

The propagation of the message signal *via* the wireless channel is very noisy, due to a number of environmental parameters. Therefore, it is recommended to model the channel before the transmission of RSSI values. To model the wireless channel, there are many proliferation models, like the free-space model, the two-ray model, LNSM (Log Normal Shadowing Model), etc. In this research, LNSM has been used because of its universal acceptance. LNSM is widely used in wireless sensor networks. The LNSM model has a feature to set the environmental parameters in real time. Hence, for any geographical as well as environmental situation, one can model the channel. The LNSM technique is discussed in detail in the next section (Section 12.3.3.2).

12.3.3.2 LNSM (Log-Normal Shadowing Model)

The RSSI signals from the wireless channel are very much prone to reflections, diffractions, and scattering. So, to tackle the problem of a noisy channel, it is important to model the wireless channels by analytical or empirical methods. In the research carried out here, the analytical method has been adopted for modeling the wireless channel. To model a wireless channel in this research, a node has been placed in the outside environment and strengthening of the signal has been observed till the pursuit and anchor nodes moved from 150 m to 1 m of distance apart. At a distance of 1 m, the signal strength should be −40 dBm in the case of ZigBee [27]. This is the first step to modeling the channel in the case of ZigBee.

To obtain the condition of the channel in an outdoor location, an experiment was conducted. Several trails have been made in aligning the pursuit node and anchor node. The pursuit node is made to move toward the anchor node from a different angle and direction. The maximum distance is kept at 150 m, although this distance can be increased. The outdoor location where the anchor node has been placed is almost flat. As already discussed, the nodes have been equipped with Xbee and SD card modules. For several trials, RSSI values have been captured and a dataset has been designed from where the estimated distance can be calculated [16]. It has been observed that

$$P_{L(d)}(dBm) = P_{L(d_0)} + 10n \log_{10}\left(\frac{d}{d_0}\right) + X_\sigma \qquad (12.1)$$

where $P_{L(d)}(dBm)$ is the path loss at a distance d (meters), $P_{L(d_0)}$ is the path loss at a distance d_0 (meters), d_0 is the reference distance, i.e., 1 meter as suggested for the Xbee module and, at 1 m, the strongest RSSI value, of −40 dBm. n is the path loss exponent calculated using the LNSM technique, and X_σ is the zero mean Gaussian random variable (in dB), with a standard deviation σ.

The RSSI values can be calculated as

$$\text{RSSI} = P_t(dBm) - P_{L(d)}(dBm) \tag{12.2}$$

where $P_t(dBm)$ is the transmitted power from the node. So, the RSSI at the pursuit node will be

$$P_{L(d)}(dBm) = P_{L(d_0)} + 10n \log_{10}\left(\frac{d}{d_0}\right) + X_\sigma \tag{12.3}$$

The value of "n" (path loss exponent) can be calculated as shown in Table 12.1. The value of n will change for different geographical and environmental conditions

12.3.3.3 Hybrid TLBO-Unilateral Algorithm

There is a need for optimization to minimize the distance error, using RSSI values to localize the trapped node. Here, the TLBO algorithm has been used for the optimization process. TLBO is a teacher–learner-based optimization algorithm. It is the best-suited algorithm as it indispensable for any algorithm-specific parameters, as with other algorithms. GA (Genetic Algorithm) is particularly dependent on some algorithm-specific parameters, like mutation probability or cross-over probability [28]; as with GA, the ABC (Artificial Bee colonial) algorithm [29] also needs some algorithm-specific parameters to configure, whereas PSO also requires its own parameters, like inertial weights, and social and cognitive parameters [30]. There are some other algorithms, such as ES (Evolution Strategy), DE (Differential Evolution), BBO (Biogeography-Based Optimizer), and AIA (Attack Intention Analysis), etc. that also requires some algorithm-specific parameters. The calibration of these parameters must be perfect, otherwise the convergence result will not be accurate and it will increase the computational efforts unnecessarily. The TLBO algorithm is one optimization technique in which one does not need to optimize such critical parameters. TLBO is divided into the teacher as well as the learner phase.

TABLE 12.1

Path Loss Exponent Values

S. No.	Path Loss Exponent (n)	Environment
1	2.0	Free space
2	1.6–1.8	Inside building (line of sight)
3	1.8	Supermarket Store
4	2.09	Conference room, with table and chairs
5	2.2	Factory
6	2–3	Inside factory (no line of sight)
7	2.8	Indoor residential area
8	2.4	Outdoor environment (this chapter)

As discussed previously, the most classical approach to locating the trapped node is trilateration and a state-of-the-art unilateral technique has also been discussed to locate the trapped node. Particularly for this research, the unilateral algorithm is proposed, as it reduces the number of nodes in the network. Now, to optimize the localization process in terms of distance error, the hybrid TLBO-unilateral algorithm has been designed. In the hybrid algorithm, the pursuit node will forage for the RSSI and, once it receives the RSSI signal of the trapped node, it teaches the pursuit nodes to obtain the strongest RSSI value at that very first step. After the first step, the learner pursuit node will become the teacher pursuit node and starts searching for the strongest RSSI signal i.e. −40 dBm. The simulation of the designed algorithm is done in SCILAB. As discussed earlier, because of the outdoor location used for the experiments, the nodes have omnidirectional RF coverage. So, to solve the localization problem, the research developed a probability- based algorithm.

The equations below describe the localization method that combines the data obtained from the RSSI values with the actual position of the sensor. After the node gets trapped, let us assume that the estimated position of the trapped node is $P_{\text{es\_pos}}$ and its exact position is $P_{\text{ex\_pos}}$. Probability theory states that

$$P_{di}(P_{\text{ex\_pos}}) = P(P_{\text{es\_pos}} \mid P_{\text{ex\_pos}}) \tag{12.4}$$

In Eq. 12.4, the P_{di} is deployment probability field function for the anchor node. Consider the RSSI value, where the estimated probable distance is

$$P_{di}(P_{\text{ex\_pos}}) = \frac{1}{2\pi\sigma_x\sigma_y} e^{\left[-\left(\frac{1}{2}\right)\left[\frac{(x_{\text{ex\_pos}}-x_{\text{es\_pos}})^2}{\sigma^2_x} + \frac{(y_{\text{ex\_pos}}-y_{\text{es\_pos}})^2}{\sigma^2_y}\right]\right]} \tag{12.5}$$

$$\sigma_x = x_m \cdot P, \sigma_y = y_m \cdot P \tag{12.6}$$

In Eq. 12.6, P is known as the error factor considered for the node. If the anchor node (ith node) receives the RSSI from the pursuit node (jth node), the distance between the two nodes can be estimated from RSSI values and the noisy estimated distance can be denoted as $d_{\text{noise}(ij)}$. The probability function will be defined for the real position $P_{\text{ex\_pos\_node\_}i}$ of the node with respect to the estimated node $\hat{P}_{\text{es\_pos\_node\_}j}$

$$P_{(i,j)}(P_{dE_i}) = P(d_{\text{noise}(ij)} \mid P_{\text{ex\_pos\_node\_}i}, \hat{P}_{\text{es\_pos\_node\_}j}) \tag{12.7}$$

Considering the noisy channel, the distance estimated is $d_{\text{noise}(ij)}$. Now, as the RSSI values have a Gaussian distribution, so the probability function can be defined as

$$P\left(d_{\text{noise}(ij)} \mid P_{\text{ex\_pos\_node\_}i}, \hat{P}_{\text{es\_pos\_node\_}j}\right) = P\left(d_{\text{noise}(ij)} \mid \partial\left(P_{\text{ex\_pos\_node\_}i}, \hat{P}_{\text{es\_pos\_node\_}j}\right)\right) \tag{12.8}$$

$$P(d_{\text{noise}(ij)} \mid P_{dE_i}, \hat{P}_{dE_j}) = \frac{1}{2\pi d_{\text{noise}(ij)}r} e^{\frac{(\partial(P_{\text{ex\_pos\_node}\_i}, \hat{P}_{\text{es\_pos\_node}\_j}) - d_{\text{noise}(ij)})^2}{2(d_{\text{noise}(ij)}.r)^2}} \tag{12.9}$$

In Eq. 12.9, r is the range error factor, which is considered to be 0.1 in our case.

Since the trapped node position and the RSSI measurement by the pursuit node are independent, the overall probability function can be combined by multiplication. Combining Eq. 12.7 and 12.9

$$F = \frac{(x_{\text{ex\_pos}} - x_{\text{es\_pos}})^2}{\sigma^2_x} + \frac{(y_{\text{ex\_pos}} - y_{\text{es\_pos}})^2}{\sigma^2_y}$$
$$+ \sum_{j \in j_i} \frac{(\partial(P_{\text{ex\_pos\_node}\_i}, \hat{P}_{\text{es\_pos\_node}\_j}) - d_{\text{noise}(ij)})^2}{2(d_{\text{noise}(ij)}.r)^2} \tag{12.10}$$

Eq. 12.10 is used to obtain the optimized distance error in a noisy channel.

12.4 EXPERIMENTAL TEST BED SET-UP IN AN OUTDOOR LOCATION AND THE DESIGN OF SENSOR NODES

Wireless sensor node design is a significant component in managing a catastrophe, that will provide reliable communication under post-disaster conditions. The design of nodes, as reported in this section, is unique in this specified area of application, and no other solution has been found to be available in the literature, to the best of our knowledge. The node consists of a Xbee radio as a transceiver, and an SD card to describe the vital information of the adjacent nodes, especially the RSSI value, along with the time log, power bank, solar panel to charge the power bank, and an LCD display. In the present study, the nodes developed to manage the disaster are classified as: anchor node (fixed nodes), trapped node (visitor node), and a pursuit node (refer to Figs. 12.8 and 12.9). The hardware architecture of all the nodes is the same except for their role in the wireless network. The fixed nodes are routers which will route the location of the visitor node to the main central server, and each has the capability of an on-board data log mechanism. The trapped node (visitor) is the end-device node, which will interact with fixed nodes (routers) while it is moving in the disaster-prone area. If, in any case, the disaster happens, the trapped node can easily be localized by a pursuit node configured as the end-node. The detailed hardware/control architecture are shown in Fig. 12.8. Figs. 12.9 and 12.10 show the fixed nodes (anchor node), the visitor node (trapped node), and the pursuit node. The whole disaster-prone area is managed by the coordinator node. The coordinator node requests the information on a timely basis from the router nodes.

Many experiments have been conducted in outdoor locations with the designed nodes in the seismic zone of Himachal Pradesh (Dharamshala to McLeod Ganj) (Fig. 12.11). The experiments for this location have been conducted, keeping in

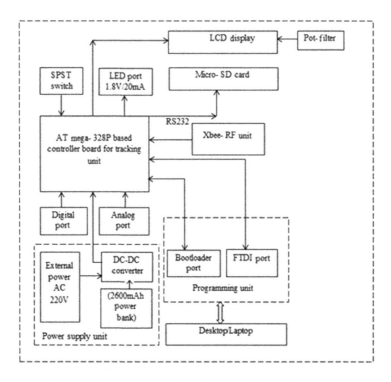

FIGURE 12.8 Detailed hardware/control architecture of Sensor Node

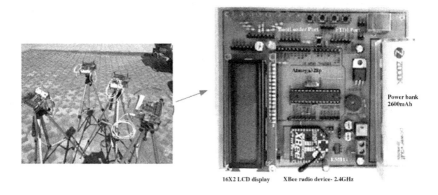

FIGURE 12.9 Fixed Nodes (Anchor Node)

mind the reliability of the network. In this section, the placement of nodes is discussed with respect to the coverage area. It is to be noted that, for a test-bed area of 150 m × 100 m, six nodes (four fixed, one pursuit, and one visitor node) are enough to perform the tests.

The exact positions of the fixed nodes have been estimated, keeping in mind the appropriate handshaking, whereas the visitor node is made to move from one fixed node to another fixed node. The visitor node can channelize and concede 20 bytes

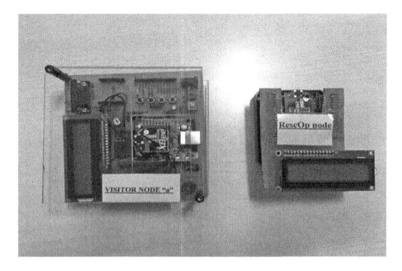

FIGURE 12.10   Visitor Node (Trapped Node) and Pursuit Node

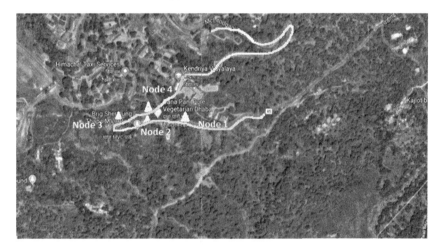

FIGURE 12.11   Node placement in OSA location (Dharamshala McLeod Ganj Road)

of data. The experiments on different test areas include the tests of RSSI (Received Signal Strength Indicator) w.r.t. the time.

12.5   VALIDATION AND RESULTS OF THE MODIFIED APPROACHES TO LOCALIZATION

12.5.1   Discussions on the Searching Pattern of Nodes

Fig. 12.12 shows the inquisitor pattern of the pursuit node, obtained using SCILAB. The virtual environment of SCILAB is designed for outdoor locations only. Fig. 12.12 presents the RSSI values in a test field. For a greater understanding, the

```
-116 -121 -119 -120 -122 -119 -117 -119 -115 -122 -118 -120 -118 -119 -120 -120 -118 -119 -121
-118 -112 -114 -112 -110 -112 -112 -115 -110 -109 -109 -111 -108 -113 -111 -112 -109 -109 -118
-117 -110 -107 -104 -105 -105 -108 -106 -107 -104 -107 -107 -108 -105 -109 -105 -103 -113 -117
-117 -111 -102  -98  -97  -97  -99  -96  -99  -99  -98 -101  -96  -99 -102  -98 -106 -109 -119
-121 -110 -105 -100  -95  -90  -90  -94  -90  -90  -90  -91  -91  -88  -92  -99 -106 -116 -116
-119 -114 -103  -98  -89  -82  -75  -79  -81  -80  -83  -84  -85  -85  -93  -99 -107 -111 -120
-118 -110 -107  -96  -89  -79  -75  -70  -67  -70  -70  -69  -69  -80  -89  -98 -102 -107 -116
-119 -110 -101  -97  -92  -80  -70  -58  -62  -55  -60  -59  -69  -79  -92  -95 -106 -109 -118
-117 -111 -105  -99  -89  -84  -71  -61  -50  -50  -50  -60  -69  -80  -92 -100 -106 -112 -117
-119 -110 -106  -94  -93  -81  -73  -63  -53  -40  -50  -60  -71  -79  -92  -99 -108 -109 -118
-123 -114 -102  -98  -91  -82  -74  -57  -50  -53  -52  -60  -71  -80  -90  -97 -104 -108 -119
-119 -115 -102  -96  -97  -80  -71  -62  -57  -55  -62  -63  -71  -82  -90  -95 -107 -112 -119
-123 -115 -108  -96  -90  -84  -73  -68  -69  -72  -69  -71  -69  -79  -86  -94 -109 -110 -117
-117 -111 -111  -95  -88  -83  -80  -83  -79  -85  -80  -82  -81  -79  -90  -97 -108 -109 -121
-119 -110 -105  -97  -91  -88  -92  -89  -92  -87  -89  -90  -90  -87  -92 -101 -107 -112 -117
-121 -110 -109  -98 -100  -97  -97  -95 -101  -95  -98 -100  -99  -99  -98  -98 -107 -111 -120
-119 -111 -109 -105 -106 -104 -106 -105 -109 -105 -108 -107 -105 -108 -104 -102 -102 -111 -120
-115 -111 -113 -109 -111 -112 -111 -110 -110 -108 -113 -112 -111 -110 -115 -110 -112 -111 -119
-115 -118 -114 -118 -115 -116 -116 -120 -119 -119 -120 -118 -120 -120 -120 -120 -119 -119 -121

-116 -121 -119 -120 -122 -119 -117 -119 -115 -122 -118 -120 -118 -119 -120 -120 -118 -119 -121
-118 -112 -114 -112 -110 -112 -112 -115 -110 -109 -109 -111 -108 -113 -111 -112 -109 -109 -118
-117 -110 -107 -104 -105 -105 -108 -106 -107 -104 -107 -107 -108 -105 -109 -105 -103 -113 -117
-117 -111 -102  -98  -97  -97  -99  -96  -99  -99  -98 -101  -96  -99 -102  -98 -106 -109 -119
-121 -110 -105 -100  -95  -90  -90  -94  -90  -90  -90  -91  -91  -88  -92  -99 -106 -116 -116
-119 -114 -103  -98  -89  -82  -75  -79  -81  -80  -83  -81  -85  -85  -93  -99 -107 -111 -120
-118 -110 -107  -96  -89  -79  -75  -70  -67  -70  -70  -69  -69  -80  -89  -98 -102 -107 -116
-119 -110 -101  -97  -92  -80  -70  -58  -62  -55  -60  -59  -69  -79  -92  -95 -106 -109 -118
-117 -111 -105  -99  -89  -84  -71  -61  -50  -50  -50  -60  -69  -80  -92 -100 -106 -112 -117
-119 -110 -106  -94  -93  -81  -73  -63  -53    *  -50  -60  -71  -79  -92  -99 -108 -109 -118
-123 -114 -102  -98  -91  -82  -74  -57  -50  -53    *  -60  -71  -80  -90  -97 -104 -108 -119
-119 -115 -102  -96  -97  -80  -71  -62  -57  -55  -62    *  -71  -82  -90  -95 -107 -112 -119
-123 -115 -108  -96  -90  -84  -73  -68  -72  -69  -71    *    *    *  -94 -109 -110 -117
-117 -111 -111  -95  -88  -83  -80  -83  -79  -85  -80  -82  -81  -79  -90    * -108    * -121
-119 -110 -105  -97  -91  -88  -92  -89  -92  -87  -89  -90  -90  -87  -92 -101    * -112    *
-121 -110 -109  -98 -100  -97  -97  -95 -101  -95  -98 -100  -99  -99  -98  -98 -107 -111 -120
-119 -111 -109 -105 -106 -104 -106 -105 -109 -105 -108 -107 -105 -108 -104 -102 -102 -111 -120
-115 -111 -113 -109 -111 -112 -111 -110 -110 -108 -113 -112 -111 -110 -115 -110 -112 -111 -119
-115 -118 -114 -118 -115 -116 -116 -120 -119 -119 -120 -118 -120 -120 -120 -120 -119 -119 -121
```

FIGURE 12.12   Path as Computed by Hybrid TLBO-Unilateral Algorithm for outdoor location

figure is divided into upper and lower halves, which clearly show the path followed by the pursuit node to locate the trapped node. As mentioned in Fig. 12.12, as soon as the pursuit node receives the RSSI signal, it will also receive the nearby RSSI signals. In the simulation environment, the pursuit node gets the −117 dBm as its first RSSI signal, and the nearby signals presented are −109 dBm, −112 dBm, −121 dBm, −120 dBm, and −111 dBm. As per the VPM protocol, the pursuit node accepts the new location i.e. −109 dBm location, as 109 dBm is the strongest signal among them all. Now, according to the hybrid TLBO and the unilateral algorithm, the pursuit node learns from the teacher to move to the next location. By this technique, the teacher identifies the best learner. Similarly, the pursuit node moves ahead and gets the new location of −107 dBm, and the localization deed continues until the target node is located. This makes the anchor node fully discernible in any outside location. The advantage of this technique is its simplicity of getting embedded into most of the electronics embedded platform. The same is tested with respect to the hardware also and the trapped node is then fully discoverable in the outdoor environment.

12.5.2   Fingerprinting Technique Validation

To validate the fingerprinting technique, a wireless network has been established in the disaster-prone area and experiments are conducted for validating the network, where RF signatures have been captured, and proper handshaking of the sensor node in the vehicle is searched for, with the fixed node deployed in the disaster-prone area. The methodology for the deployment of such a network has already been discussed in Section 12.3. As defined, the vehicles with their respective sensor nodes were made to move, and respective RF signatures were detected. The dereliction ratio of handshaking and capturing the RF signatures was much lower. The vehicle was made to pass the disaster-prone area at different speeds, i.e., from 20 km/h to 50 km/h. The packet loss was more when vehicle speed was more than 45 km/h and less when the vehicle speed was less than 40 km/h. The terrain was hilly, so the vehicle maximum speed was not more than 50 km/h. There is a loss of one or two packets only at speeds of more than 45 km/h. The results from handshaking of the nodes were also observed, when the vehicle is moving from location 1 to location 2, and to other location. The handshaking between the nodes was easy and smooth. RF values were captured in the fixed nodes. In Fig. 12.13, the RF signature was captured by fixed nodes of location 1, with the vehicle carrying the movable node. As captured, the signal varies from −90 dBm to −48 dBm. Similarly, when moving the MN1 away from the fixed node of location 1, then the signal was captured, and it varied from −40 dBm to −90 dBm, as shown in Fig. 12.14. The same scenario was observed for other locations too, as shown in location 2 and location 3 in Fig. 12.15 and Fig 12.16, respectively. In Fig. 12.17 it was observed that, if the speed of the vehicle exceeded 45 km/h, then packet loss took place. The packet loss is easily observed in Fig. 12.18. The results show the exceptional behavior of reliability of the fixed as well as the movable node.

12.6 CONCLUSION AND FUTURE SCOPES

To achieve high-quality mobility in a disaster-prone area, sensor networks are one of the most promising technologies that automotive industries can adopt. This

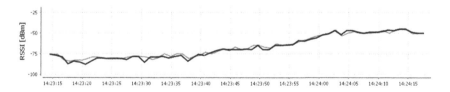

FIGURE 12.13   RF signature by node in location 1 (vehicle carrying node moves toward fixed node in location 1)

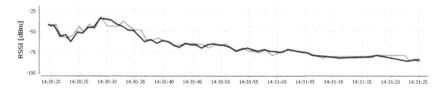

FIGURE 12.14   RF signature by node in location 1 (vehicle carrying node moves away from fixed node in location 1)

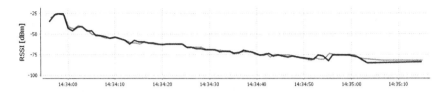

FIGURE 12.15   RF signature by node in location 2 (vehicle carrying node moves away from node in location 2)

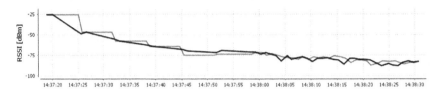

FIGURE 12.16   RF signature by node in location 3 (vehicle carrying node moves away from node in location 3)

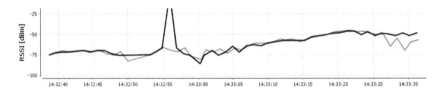

FIGURE 12.17   RF signature by node in location 4 with speed of 48 km/h

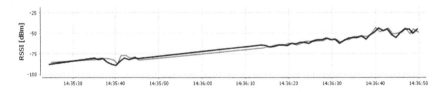

FIGURE 12.18   RF signature by node in location 4 with speed of 25 km/h

chapter discussed the design, development, and deployment of a sensor network for the mobility of vehicles in disaster-prone areas. The proposed sensor network has a capability of tracking trapped vehicles. The algorithms, such as location fingerprinting and unilateral techniques, were used to localize the vehicle, and these two algorithms are the backbone of the proposed network. The location fingerprinting technique is used to design the radio map by capturing the RF signatures used to track the vehicle in real time for pre-disaster as well as post-disaster situations. Many trails were conducted to validate the algorithm in simulation as well as in real time. The study was conducted over a limited area; in the future, we will extend the study to cover larger areas too and to cover other geographical areas. In future, we will also consider energy consumption in our study.

However, it must be noted that the developed system has not been designed for all disasters and for all geographical regions. This must be kept in mind with respect to the broad spectrum of the current study, thereby making it a larger task. However, should floods be considered the focal point of the node design, the details, especially its structure, will greatly change. In that case, it will be a completely different research study where one has to think of a floating node concept.

The proposed system improves the sustainability quotient, and also has a negligible impact on the environment. At the same time, its working is affected little by changes in the environmental conditions in the proposed areas. No harmful products are emitted into the environment. The system is not dependent on any infrastructure which might otherwise get destroyed in case of a disaster. The smaller occupied area also helps to maintain the ecological conservation of the geographical region. Even the maintenance and up-grading of the nodes, as well as the change in node locations, are not major tasks. The feature of reliable communication strongly favors sustainable development.

REFERENCES

1. Brundtland, G. H. Our common future—Call for action. *Environmental Conservation*, 1987. **14**(4): pp. 291–294.
2. Mukherjee, S., *Analytical Modeling of Heterogeneous Cellular Networks*. 2014. Cambridge University Press.
3. Usman, A. and S.H. Shami, Evolution of communication technologies for smart grid applications. *Renewable and Sustainable Energy Reviews*, 2013. **19**: pp. 191–199.
4. Berke, P.R., J. Kartez, and D. Wenger, Recovery after disaster: achieving sustainable development, mitigation and equity. *Disasters*, 1993. **17**(2): pp. 93–109.
5. Pearce, L., Disaster management and community planning, and public participation: how to achieve sustainable hazard mitigation. *Natural Hazards*, 2003. **28**(2–3): pp. 211–228.
6. O'Brien, G., et al., Climate change and disaster management. *Disasters*, 2006. **30**(1): pp. 64–80.
7. Bello, O.M. and Y.A. Aina, Satellite remote sensing as a tool in disaster management and sustainable development: towards a synergistic approach. *Procedia-Social and Behavioral Sciences*, 2014. **120**: pp. 365–373.
8. Shaw, R., F. Mallick, and A. Islam, *Disaster Risk Reduction Approaches in Bangladesh*. 2013. Springer.
9. Sahay, B., N.V.C. Menon, and S. Gupta, Humanitarian logistics and disaster management: the role of different stakeholders, in *Managing Humanitarian Logistics*. 2016, Springer. pp. 3–21.
10. Ooi, G.L., et al., Near real-time landslide monitoring with the smart soil particles. *Japanese Geotechnical Society Special Publication*, 2016. **2**(28): pp. 1031–1034.
11. Gioia, E., et al., Application of a process-based shallow landslide hazard model over a broad area in Central Italy. *Landslides*, 2016. **13**(5): pp. 1197–1214.
12. Wu, C.-I., et al., An intelligent slope disaster prediction and monitoring system based on WSN and ANP. *Expert Systems with Applications*, 2014. **41**(10): pp. 4554–4562.
13. Kohvakka, M., et al. Performance analysis of IEEE 802.15. 4 and ZigBee for large-scale wireless sensor network applications. in *Proceedings of the 3rd ACM International Workshop on Performance Evaluation of Wireless ad hoc, Sensor and Ubiquitous Networks*. 2006. ACM.

14. Ali, A., et al., Efficient predictive monitoring of wireless sensor networks. *International Journal of Autonomous and Adaptive Communications Systems*, 2012. **5**(3): pp. 233–254.

15. Hartenstein, H. and L. Laberteaux, A tutorial survey on vehicular *ad hoc* networks. *IEEE Communications Magazine*, 2008. **46**(6): 164–171.

16. Gharghan, S.K., et al., Accurate wireless sensor localization technique based on hybrid PSO-ANN algorithm for indoor and outdoor track cycling. *IEEE Sensors Journal*, 2016. **16**(2): pp. 529–541.

17. Nedjati, A., B. Vizvari, and G. Izbirak, Post-earthquake response by small UAV helicopters. *Natural Hazards*, 2016. **80**(3): pp. 1669–1688.

18. Zhou, Y., et al., Multi-UAV-aided networks: aerial-ground cooperative vehicular networking architecture. *Ieee Vehicular Technology Magazine*, 2015. **10**(4): pp. 36–44.

19. Tuna, G., B. Nefzi, and G. Conte, Unmanned aerial vehicle-aided communications system for disaster recovery. *Journal of Network and Computer Applications*, 2014. **41**: pp. 27–36.

20. Rusu, C.V. and H.-S. Ahn. Optimal network localization by particle swarm optimization. in *Intelligent Control (ISIC), 2011 IEEE International Symposium on*. 2011. IEEE.

21. Payal, A., C.S. Rai, and B.R. Reddy, Analysis of some feedforward artificial neural network training algorithms for developing localization framework in wireless sensor networks. *Wireless Personal Communications*, 2015. **82**(4): pp. 2519–2536.

22. Thongpul, K., N. Jindapetch, and W. Teerapakajorndet. A neural network based optimization for wireless sensor node position estimation in industrial environments. in *Electrical Engineering/Electronics Computer Telecommunications and Information Technology (ECTI-CON), 2010 International Conference on*. 2010. IEEE.

23. Nerguizian, C. and V. Nerguizian. Indoor fingerprinting geolocation using wavelet-based features extracted from the channel impulse response in conjunction with an artificial neural network. in *Industrial Electronics, 2007. ISIE 2007. IEEE International Symposium on*. 2007. IEEE.

24. Rahman, M.S., Y. Park, and K.-D. Kim, RSS-based indoor localization algorithm for wireless sensor network using generalized regression neural network. *Arabian Journal for Science and Engineering*, 2012. **37**(4): pp. 1043–1053.

25. Kaundal, V., P. Sharma, and M. Prateek, Wireless sensor node localization based on LNSM and hybrid TLBO-unilateral technique for outdoor location. *International Journal of Electronics and Telecommunications*, 2017. **63**(4): pp. 389–397.

26. Rao, R.V., Teaching-learning-based optimization algorithm, in *Teaching Learning Based Optimization Algorithm*. 2016, Springer. pp. 9–39.

27. ZigBee Specification, ZigBee Alliance. ZigBee Document 053474r06, Version, 2006. **1**.

28. Elsayed, S.M., R.A. Sarker, and D.L. Essam, A new genetic algorithm for solving optimization problems. *Engineering Applications of Artificial Intelligence*, 2014. **27**: pp. 57–69.

29. Karaboga, D., et al., A comprehensive survey: artificial bee colony (ABC) algorithm and applications. *Artificial Intelligence Review*, 2014. **42**(1): pp. 21–57.

30. Cao, C., Q. Ni, and X. Yin, Comparison of particle swarm optimization algorithms in wireless sensor network node localization. in *Systems, Man and Cybernetics (SMC), 2014 IEEE International Conference on*. 2014. IEEE. pp. 252–257.

13 Offline Payment System for Public Transport

Bhavesh Praveen, Ratanjot Singh, and Swarnalatha P

CONTENTS

13.1 INTRODUCTION

Digital currency has revolutionised payment methods, with respect to methods such as Paytm wallet, BHIM UPI, net-banking and mobile banking. This has made us think about leveraging such existing trends but use it for the greater good of the common people who regularly use hard cash for the most mundane daily tasks, which includes commuting by bus. India has many people commuting by bus every single day, and only the conductor knows how hard it is to keep track of people, who has got the tickets, and how much change he owes the passengers. Regular passengers take a monthly pass, others keep digging in their pockets, bags, and wallets for change [1]. Enter OOPS (offline-offline payment system). Imagine an offline PayTm wallet but for bus commuting; that represents a version of our project. Here, the passenger, when signing up for the scheme, pays an appropriate amount (let's say Rs.100) as deposit and puts up an additional amount for whatever the user needs for commuting, using the same logic as Paytm wallet, in the OOPS virtual wallet [2]. When riding the bus, the passenger opens our mobile app ("passenger-app") and asks the bus conductor for a ticket from his/her starting point to the destination. The conductor uses our conductor-app on the device to make connection with the passenger with audioQR (near sound data transfer) technology, which is what Google Pay uses. The passenger transfers the money from the virtual wallet to the conductor's account; if the transfer is successful, the conductor sends the ticket confirmation to the passenger along with an updated balance on the passenger's wallet. The transaction details include: unique transaction id (number indicative of bus route, bus number, and bus conductor), unique passenger id, starting point, destination, the bus tariff, balance to be updated on the passenger wallet, and finally a time stamp of the transaction. Up to this point, the entire procedure is carried out offline. The process requires no internet connectivity, only the need for two android phones, one having the passenger-app and the other having the conductor-app [3]. Near Sound Data Transfer (NSDT) is a sound-based portable exchange innovation, created and protected by Tagattitude since 2005. NSDT utilizes the idea of one-time password (OTP), which is sent through the sound channel of a cell phone, to make an electronic mark to perform secure exchanges. This technology is compatible with all smartphones used in the world because it uses the device's audio channel (speaker and microphone) for the transaction. The primary use of NSDT is for mobile banking transactions through the mobile platform Tagpay [4, 5]. It is a way of enabling authentication on websites by securely "opening doors".

13.2   STATE OF THE ART

This is an offline payment app. The basic working of this system is that payment can be made offline, i.e., without internet connection. This is exhibited when the payments are queued, while not connected to the internet. Given below in Table 13.1 is a literature survey of the offline payment trend in India *vs* abroad.

Bitcoin, which is an appropriated method of exchange wherein a bank plays no role, is an effective cryptocurrency. The preparation expenses to the banks are reduced, which is a helpful and favorable position. On the other hand, its exchanges are open, and sensitive exchanged information can be followed by other people. This is a significant disadvantage [6]. Likewise, the size of the e-coin steadily increases with every exchange and consequently limits the number of clients.

Some e-payment systems proposed have security imperfections: an assault on the detection of double spending, possible fashioning of the expiration date of a

---

TABLE 13.1

Offline Payment Trends in India

National	International
Google Pay, Amazon Pay plan to join with local developers for offline payment.	It is still more inclined toward online payment.
Cashless payment is growing steadily.	Cashless payment is well established.
Some of the big offline payment gateways include PayTM	Some of the big offline payment gateways include Stripe, PayPal.
Offline payment is used less, though it can be a necessity.	Offline payment is considered to be only an added bonus.
Offline payment exists through e-wallets.	Offline payment exists through e-wallets.

legitimate e-coin, and cheating on the exchange convention. In these systems, an intruder can pull back coins without inputting their real identity, and forge a legitimate coin utilizing the homomorphic property of modular operations [7, 8]. Thus, most of the recent schemes proposed contain some major flaws, resulting in security issues, and, hence, they are not deployable. None of them have been accepted globally (Table 13.1).

13.3 PROBLEM STATEMENT

With OOPS, we can make fast transactions because payments are made offline, and the details are later synced with our server online. We can help with live bus tracking, which is a superpower every Indian aspires to have. If the conductor has internet connectivity, we would be pushing bus ticket transactions to the cloud ("firebase" in our scenario); otherwise, transactions will queue the next time the conductor has internet access to post them as batches [9]. At the end of every bus ride or at the end of every day, the conductor has to upload the bus ticket transactions. This is essential and is a convention followed in the ordinary ticketing system as well. If the conductor has prolonged internet connectivity, as mentioned in the above scenario, we can also do bus live tracking, which usually every passenger in the world dreams of.

13.4 PROPOSED SYSTEM

In our system, the user has to create an account with the help of the mobile number after he/she installs the application ("app"). The mobile number is verified with the help of an OTP. Next, the user has to lodge money into the OOPS virtual wallet, with the help of a debit/credit card. To purchase the ticket, the passenger has to select "buy ticket", at which point audio signals are sent from the phone. The conductor has to receive the audio signal from the passenger's phone. The details of the ticket received, namely the amount and ID, are visible on the conductor's screen. The passenger can retry if the transaction fails for any reason [10].

The passenger, conductor, and the bus depot management interact with the system directly. All are given initial access to the internet, with the bus depot being an exception by having an internet connection at all times. The communication between the bus passenger and bus conductor is direct and both parties interact directly with the software through the developed interface. The bus service management accesses the entire system directly. Fig. 13.1 shows the appropriate Process Flow Diagram

13.4.1 SYSTEM ARCHITECTURE

We used multi-modular MVVM architecture for the Android app. There are a lot of other software architectures that exist, like MVC (Model View Controller), MVP (Model View Presenter), MVVM (Model View ViewModel), MVI (Model View Intent), VIPER, MOBIUS, and MVRx (Model View Reactive extensions), etc. But the most popular ones are MVP and MVVM. Since the start of Android phone development, there has been no specific architecture or guidelines given by anybody,

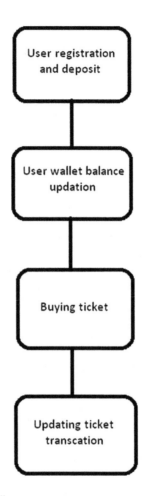

FIGURE 13.1   Process flow diagram

so everyone could choose whatever system suited them. This resulted in inconsistencies among projects in various organizations, which put the burden on developers to learn many different architectures. But to maintain standards, people started using MVP architecture [11]. Soon, people realized that this architecture cannot scale so Google themselves released a guide to app architecture. This architecture was based on MVVM architecture, with the simple addition of a repository layer.

MVP uses a more generalized version of MVVM but, in Android, we have multiple sources of data, like local SQLite Database and remote REST API source. To manage this multiple data source, we have to use a layer called Repository. Repository depends both on local DataAccessObjects (DAOs) and remote WebServices to provide data to ViewModels, which Views observe. Fig. 13.2 shows the System Architecture.

In real life, the repository first returns a whole table as an observable variable, which will be empty at first. After this, the repository immediately makes a REST

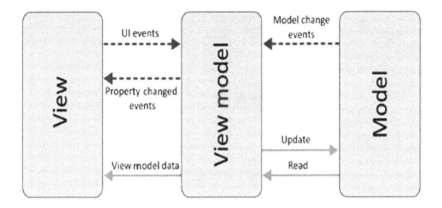

FIGURE 13.2 System architecture

API call to fetch data and insert them into the database and now, since UI (View) is observing the database, it will now show the updated data. This is a simple example of reactive programming [12].

13.4.2 INTERNAL ARCHITECTURE

The modules defined are as follows:

- Models
- Local
- Remote
- Repository
- Common
- Feature-Authentication
- Feature-Passenger
- Feature-Conductor
- Core/App

After the passenger taps the "Buy Ticket" button, the app generates audio signals using the device's speaker.. Fig. 13.3 shows the corresponding Internal Architecture

The app will support recharging the virtual wallet through a number of payment methods, such as credit or debit cards issued by almost all the banks in our country. Only a person with a verified mobile number is allowed to make transactions (payments/lodgments) though this app, for security reasons. In case of any money thefts, the owner of the mobile with that particular mobile number will be held responsible.

After the end of each working day, in the bus depot, the summary of the transactions is uploaded online, for calculating the profit, and for maintenance of the record. The server in the bus depot is connected with a cloud service for database management and processing needs. In this way, we have eliminated the need for a local database.

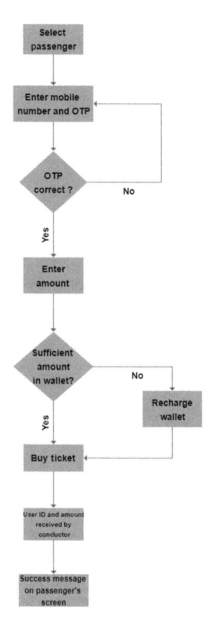

FIGURE 13.3   Internal architecture

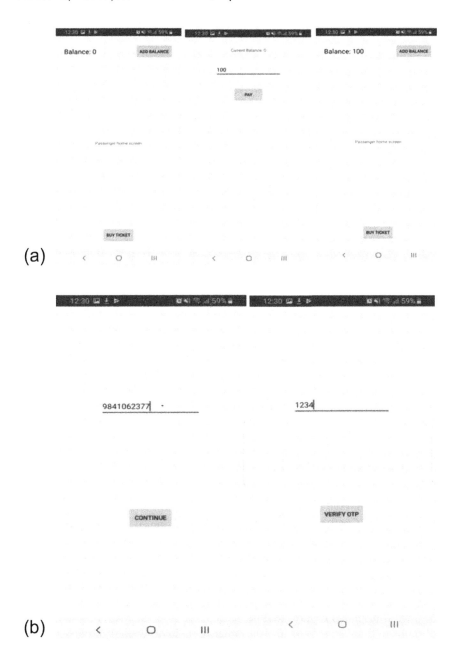

FIGURE 13.4 (a)–(c) Working demonstration

(c)

FIGURE 13.4   (Continued)

13.5   RESULTS AND DISCUSSION

The application we designed was perfectly able to send and receive money from the passenger side and the conductor side, respectively. The audio signals worked correctly as expected and, even when on noisy public transport, like a local bus, the app was able to send and receive the audio signals with correct results.

The concept of the one-time password (OTP) ensures that only genuine customers can transact through this app, and no person is able to impersonate someone with his mobile number. Since the medium of communication is an audio channel, there is no need for special encryption/decryption mechanism for safety. The local transaction ensures that no hacker can intercept and/or steal any information from the system. The non-dependence on internet connectivity is a major plus point as it can be used in remote areas where network signals can be intermittent and/or very weak. Also, analysis of the trends in ticket purchase can be carried out on the data on the central database, which is updated at the depot every day after the working day has ended. Fig. 13.4 (a)-(c) show a working demonstration of the app.

13.6 FUTURE SCOPE

Efforts are already being made to improve upon this initial model. With the use of Big Data technologies at the server end, it would become a lot faster and more secure to store the cumulative data. This will be helpful in finding variations and trends in the travel sector. More transport facilities could be deployed on busy routes, whereas, on routes where travel is less frequent, the frequency of the vehicles could be reduced. Machine learning is a domain in which we can focus our efforts for providing smart solutions, using facial recognition, combined with other biometrics for additional security.

There is always scope for improvement in any of the systems built. Even if the slightest of changes can enhance the system, the change should be considered; the security mechanism and the audio speed could be further improved, for example, if we use a different medium of communication. Security mechanisms, which are standard and widely accepted, can be used in the wireless communication, to achieve greater security. Newer security issues are arising each day, and maintenance of the security of the system is considered a challenging task these days.

13.7 CONCLUSION

In our paper, we have proposed a model for an offline payment system which is better than the payment apps which are on the market today, because our application does not depend solely upon internet connection. In a country like ours, where a majority of the population lives in remote areas, having an internet connection all the time is a very challenging task. So, our application can work well in those areas. Also, during natural calamities, when the communication systems are completely shut down, this application can turn out to be a life saver in terms of transport; it could also be adapted to allow the purchase of important items. Having a cloud service at the back end enables us to have data stored and retrieved faster, thus eliminating the need for having a local database and a server maintenance person at all times. Cloud storage also keeps the data more secure. Looking at the amount of data thefts and other security attacks, our system will be able to cater to the changing security needs as our means of data communication, namely audio waves, is very secure.

REFERENCES

1. Mrs. Dhanalakshmi Komirisetty, Mr. B. Sarath Simha. A Study on Paytm's Growth in India as A Digital Payment Platform. *International Journal of Research and Analytical Reviews (IJRAR)* 5(4), December 2018.
2. Abdullah Al Rahat, Kazi Md. Rokibul Alam, R. Tahsin, G.G. Md. Nawaz Ali. All Offline Electronic Payment System Based on an Untraceable Blind Signature Scheme. *KSII Transactions on Internet and Information Systems* 11(5): 2628–2645, January 2017.
3. Karamjeet Kaur, Dr. Ashutosh Pathak. E-Payment System on E-Commerce in India. *International Journal of Engineering Research and Applications* 5(2): part 1, February 2015.

4. P.S.V.S. Sridhar, T.V.S. Rohitkumar, G. Dharani, V.B.S. Akhila. Introducing Secured Offline Payments using FRoDO. *International Journal of Innovative Technology and Exploring Engineering (IJITEE)* 8(7s), May 2019.

5. Bazeem Ismaeil Khan, Zubair Ahmed Shaikh. Secure Device for Offline Micro Payment. *International Journal of Emerging Trends & Technology in Computer Science* 24(4): 33–37, March 2017.

6. T. Takahashi, Akira Otsuka. Secure Offline Payments in Bitcoin, Institute of Information Security, Yokohama, Japan. Available at https://fc19.ifca.ai/wtsc/Secu reOfflinePayments.pdf.

7. Siamak Solat. Security of Electronic Payment Systems: A Comprehensive Survey. Available at https://arxiv.org/ftp/arxiv/papers/1701/1701.04556.pdf.

8. R. Bargavi, Dr. L. Jaba Sheela. Fraud Resilient Device Offline Micro-Payments using Bit-Exchange Algorithms. *International Journal of Engineering and Computer Science* 6(3): 20699–20704, March 2017.

9. P. Shana Jebin. PayOff: An Approach for Offline microPayments. *International Journal of Innovative Research in Computer and Communication Engineering* 5(3), March 2017.

10. Ki-Woong Park, Sung Hoon Baek. OPERA: A Complete Offline and Anonymous Digital Cash Transaction System with a One-Time Readable Memory. *IEICE TRANSACTIONS on Information and Systems* 100(10): 2348–2356, October 2017.

11. Prasad A. Naik, Kay Peters. A Hierarchical Marketing Communications Model of Online and Offline Media Synergies. *Journal of Interactive Marketing* 23(2009): 288–299, November 2009.

12. Dennis Abrazhevich. Electronic Payment Systems: A User-Centered Perspective and Interaction Design. Available at https://pure.tue.nl/ws/files/2396269/200411085.pdf.

14 Microbial Fuel Cell
A Source of Bioelectricity Production

Gagandeep Kaur, Akhil Gupta, and Jaspreet Kaur

CONTENTS

14.1 INTRODUCTION

In addition to the considerable (and increasing) demand for energy in rural and urban Indian communities, the trend is gradually shifting from non-renewable energy sources to renewable ones, the latter being acknowledged as effective alternatives for generating energy for distribution to consumers. Agriculture-, forest-, and livestock-based biowastes or by-products are called biomass or bioresidues, and are available in large quantities in India [1, 2]. Biomass-based energy generation is popular in rural areas due to infrastructural constraints to delivery *via* conventional sources. Bioenergy has merit as it is renewable and extractable from organic matter by utilizing simple and economical techniques, and processes of anaerobic digestion (AD), yielding high levels of practically usable biogas [3–5]. Biogas is a recognized eco-friendly energy source, with the main components being methane (60–70%) and carbon dioxide (30–40%) [6–8].

Another innovative technique, developed in recent years to utilize available biomass for energy generation, is the Microbial Fuel Cell (MFC). MFCs are currently under intensive research and researchers have been able to obtain a maximum power density of 3600 mW/m² [11] with a glucose-fed substrate, using commonly available raw biomass constituents. A typical MFC is a bioreactor which converts chemical energy, existing in bioconvertible substrates, directly into electricity by the action of specific microorganisms which facilitate the conversion of substrate directly into electrons [9–11].

14.1.1 MICROBIAL FUEL CELLS

Energy consumption across the globe has increased exponentially during the first decade of the 21st century and is continuing to do so. To meet the ever-increasing energy demand, there is a need to identify more and sustainable feasible sources of energy. Indiscriminate exploitation of fossil fuels to meet demand has posed a threat to biological life on the planet *via* its secondary effects of global warming and environmental pollution [11, 12]. The dire need for alternatives to fossil fuels has encouraged researchers to seek alternative sources of power which can be harnessed by utilizing modern tools of technology developed in recent years. Proper and opti-mised use of renewable energy resources may be an answer to this serious problem. An extensive range of energy solutions have been explored by researchers, because any one of the presently available alternatives is unlikely to replace fossil fuels. As a consequence of these efforts, one of the recently proposed alternatives is energy derived from fuel cells, utilising microbial digestion of biomass [13, 14].

An electrochemical engine, which converts the existing energy of chemical bonds into electricity, is called a fuel cell [15, 16]. Being a green source of energy, this option seems attractive, as the energy obtained thereof is both renewable and envi-ronmentally friendly. Fuel cells utilizing biological material for power generation involve enzymatic catalysis of ingredients in an electrolysis chamber. Biological fuel cells are capable of directly transforming chemical energy to electrical energy by way of electrochemical reactions. There are two types of biological fuel cells, namely Microbial Fuel Cells (MFCs) and Enzymatic Fuel Cells (EFCs). If biological fuel cells are using biomass to act as substrate for bioelectricity production, then they may be named biomass fuel cells [17].

MFCs are novel devices that use a bacterial community as the biocatalyst for the oxidation of organic (or inorganic) matter to generate current [18]. A biopotential, developed between the bacterial metabolism and the substrate, leads to the gener-ation of bioelectricity in MFCs. Anaerobic conditions are necessary in the anode chamber as oxygen will hinder the production of electricity, so that a pragmatic arrangement must be designed, in which the bacteria are separated from oxygen [11, 17–19].

MFC is an impressive technology, with the capability to digest a wide range of substrates with bacteria to generate bioelectricity, despite the fact that power lev-els are low. It is mostly preferred for sustainable long-term power applications [11, 20–22]. As normal fuel cell (FC), being a conventional energy resource, energize the distributed generation (DG) units of power system. Distributed Generations (DGs), a term commonly used for small-scale generations, offer solution to many of new energy generation challenges. DG is an electric power generation source connected directly to the distribution network or on the customer side of the meter, having gen-eration from 'a few kilowatts up to 50MW. Similarly, with high-power generation capabilities, MFC may act as a source for distributed electricity generations. Fig. 14.1 shows a schematic of the basic components of a double-chamber MFC.

14.1.2 MECHANISM OF MICROBIAL FUEL CELLS

MFCs utilize microbes as the catalysts to oxidize organic matter in these bio-electrochemical devices, to generate current. An MFC unit, as shown in Fig. 14.1, is a double chamber having an anodic as well as a cathodic chamber, the two being separated by a semi-permeable membrane, generally known as a proton-exchange membrane (PEM). In the anodic chamber, the microflora results in the generation of protons and electrons *via* oxidation of organic matter in an anaerobic environment, generating carbon dioxide and other compounds as final products. The protons travel to the cathode chamber through the membrane and the movement of electrons generated in the process is facilitated *via* an external circuit, where electrons are transmitted to the cathodic chamber. In the cathode chamber, protons and electrons react, along with the parallel reduction of oxygen to water. Therefore, bioelectricity is generated in an MFC by bacterial metabolism, due to the development of biopotential. MFCs are gaining consideration due to their capability to use a variety of biodegradable substrates under mild conditions. An air–cathode MFC, shown in Fig. 14.2, is a single-chamber MFC, in which the anode is placed in the anodic chamber where organic matter is present. The cathode is pasted outside the anodic chamber, separated by the PEM and exposed to the air. The working principle of the air–cathode MFC is the same as for the double-chamber MFC.

MFC performance depends mainly on several important factors, such as the system configuration, the nature of the organic matter, the bacterial species, the electrode material and surface area, type of catholyte, operating conditions, rate of oxidation in the anodic chamber, electron shuttle from the anodic chamber to the surface of the anode, the way of supply organic matter into MFC, consumption rate

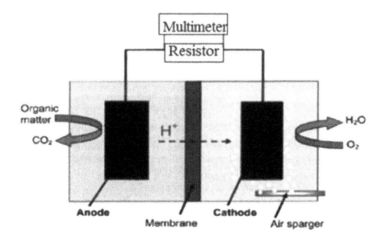

FIGURE 14.1 Schematic of the basic components of a double-chamber MFC.

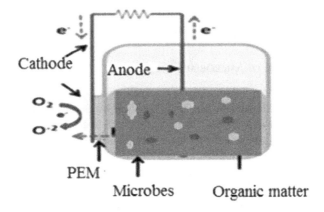

FIGURE 14.2   Schematic of an air–cathode single-chamber MFC

in the cathode chamber, and the permeability of the PEM [11, 22]. Anaerobic conditions are essential for the anodic compartment, as the configurations are designed for an oxygen-free region [11, 17, 18]. A continuous supply of biological raw material at regular intervals is necessary to ensure a steady generation of electrical energy [23, 24]. The chemical reactions taking place in anode and cathode chambers for an organic substrate are as follows:

$$C_6H_{12}O_6 + 6H_2O = 6CO_2 + 24H^+ + 24e^- \quad \left(\text{Anode chamber}\right)$$

$$24e^- + 24H^+ + 6O_2 = 12H_2O \quad \left(\text{Cathode chamber}\right)$$

$$C_6H_{12}O_6 + 6H_2O + 6O_2 = 6CO_2 + 12H_2O \quad \left(\text{Net MFC reaction}\right)$$

In addition to the generation of bioelectricity, the end products are carbon dioxide and water. About 24 electrons participate in the flow of current, with bacteria acting as the catalysts to activate the chemical reactions.

14.1.3   MICROBIAL FUEL CELL TECHNOLOGY AND ADVANCES

Potter (1911) [15] introduced the concept of MFC and reported that any physiological process, accompanied by chemical changes, involves a related electrical change. The breakdown of organic compounds by micro-organisms is accompanied by the release of electrical energy. With the action of microorganisms, the electrical effects are introduced and are influenced by temperature, the number of active bacteria, and the concentration of the nutrient medium. These effects are limited by the temperature favourable for the microorganisms and for protoplasmic activity. The maximum recorded voltage from an MFC was 0.3–0.5 V.

Based on this concept, Davis et al. (1962) [19] experimented to determine the role of microbes and hydrocarbons in the generation of electrical energy. The addition of glucose oxidase or microbes to a solution of glucose resulted in electrical output. In the absence of oxygen, biological dehydrogenation took place and it was considered that, with a hydrogen ionisation reaction, a wire could link oxygen with microbial dehydrogenations. The electrons transferred through the semi-permeable membrane produced hydroxyl ions at the oxygen electrode and reacted with hydrogen ions to complete the cyclic process. Experimental findings have shown that addition of methylene blue increased the open circuit voltage (OCV) from 80 to 180 mV and from 50 to 100 mV, maintained under 1000 ohms load. Similarly, addition of the gut bacterium *Escherichia coli* increased OCV from 150 to 625 mV and to 500 mV under a load of 1000 ohms. Addition of potassium ferricyanide resulted in only a slight increase in current.

Berk (1964) [20] reported a study of the interaction between electrode material and photosynthetic microorganisms. A sandblasted platinum electrode, on which marine algae were growing, generated a current density of 4.3 $\mu A/cm^2$ with a 0.6 V potential. Appropriate combinations of bioelectrodes, using *Rhodospirillum rubrum* (a bacterium which is photosynthetic under anaerobic conditions) with malate, have shown the capability for light-dependent production of electrical energy.

In early2000, rigorous research had started to increase the generation capabilities of MFCs and Steele et al. (2001) [16] presented data that fuel cells operate at high efficiency with low levels of pollutants in the production of electrical energy. The vital issues relating to fuel cell technology, such as alternative materials for the stacking of fuel cells and optimal selection of fuels for MFCs, were discussed. Present cells use traditional materials but commercialization studies and cost analysis have uncovered the limitations of these materials.

Logan et al (2006) [18] reported that research into MFCs was developing swiftly but lacked the methods of evaluation for system performance. Researchers were facing technical problems in comparing the performances of MFCs on an appropriate basis with conventional electricity generation systems. MFC construction and performance studies require information on microbiology, materials electrochemistry, and fundamentals of engineering. Performances of MFCs constructed in different configurations and from different materials were being analyzed by standard polarization curves.

You et al. (2006) [25] reported that MFC is a novel bioprocess, producing electrical energy from organic matter. A peak value of power density of 115.60 mW/m^2 was obtained in two-chamber MFCs, with permanganate as the cathodic electron acceptor as compared to hexa-cyanoferrate and oxygen, with power densities of 25.62 mW/m^2 and 10.2 mW/m^2, respectively. In comparison to double-chambered MFC, a bushing MFC (a different MFC reactor design), using permanganate, achieved an unparalleled maximum power output of 3986.72 mW/m^2. This study has presented permanganate as an effective electron acceptor to augment MFC efficiency.

Lovely (2006) [12] reported that, though the technology for MFC has been established, there has been less development in practical usage than would have been expected. Sediment MFC has shown the practical application, for feasibility studies, of electricity generation in remote areas. MFC has the capability to treat a range of organic wastes to make MFCs a feasible self-sustaining source for electricity generation. With recent developments, power output of MFC has increased but it still needs optimization of parameters to achieve large-scale electricity production.

14.1.4 RECENT DEVELOPMENTS

Fornero et al. (2010) [21] explored the feasibility of MFC technology for wastewater treatment. They discussed the problems occurring with respect to generation of current from complex wastewater due to different types of microbes in the microbial community. This diversity led to undefined microbial communities, low coulombic efficiencies, slow kinetic rate, and non-linear power density increase during scaling. To analyze these parameters in MFCs, comparison between studies are difficult, due to the use of different electrode materials, membranes, substrates, bacterial communities, configurations, electrode conductivities, electron transfer rate, temperatures, and pH levels.

Franks and Nevin (2010) [14] reported that MFCs have the capability to treat a broad range of organic substrates for the generation of electrical energy. The intensity of research in this field has increased many fold, to assess organic waste as a sustainable energy producer. Sediment MFCs have been successful to provide current for low-power applications. For advances in MFC technology, knowledge of the limitations of the technology and of the behavior of microbes is required. Several researchers consider the greatest achievement of MFCs is the ability to treat organic waste and to degrade toxic wastes rather than electricity generation. More emphasis is required on the understanding of the microbial process in MFCs for further development of practical applications.

Gupta et al. (2011) [17] reviewed biofuel cells and their classification into either microbial fuel cells or enzymatic fuel cells. The main focus was on performance and developments made in MFCs, as here the challenge is to achieve the correct blend of biological parameters with electrodes. Researchers have confirmed the capability of MFC technology for low-power generation and the means to degrade organic waste and toxic chemicals. The authors reviewed the research on MFC in terms of electrodes, performance evaluation methods, and environmental treatments. Various approaches to overcoming existing challenges were reviewed. This potentially high-impact technology has applications mainly in the field of energy generation from organic materials, as well as in clinical research and medical sciences.

Cheng et al. (2011) [26] stated that, to meet the challenge of the scaling-up of MFCs, the importance of various significant parameters, like the surface area of the electrodes, reactor geometry, substrate conductivity and concentration, is essential. Substrate conductivity affects the cathode, whereas substrate concentration affects the anode significantly. Using wastewater as the substrate for scaling-up MFCs, the most important deciding factor is the cathode-specific surface area. In favor of power generation, volumetric power density increases linearly with cathode surface area, but substrate strength and conductivity need to be high. Higher volumetric densities are possible with reactor configurations of smaller liquid volumes and closer electrode spacings. Studies concluded that the most essential feature for scaling-up of MFCs is cathode surface area.

Rahimnejad et al. (2012) [22] carried out studies on a novel stack of four MFCs of a bio-degradable material at continuous mode for production of clean and sustainable energy. An active biocatalyst, the yeast *Saccharomyces cerevisiae*, was added to enhance the power generation capability. In pure glucose substrate, addition of Natural Red as a mediator in the anode chamber and potassium permanganate as an oxidizing agent in the cathode achieved a maximum current of 6447 mA/m^2 and power generation of 2003 mW/m^2. Electrical performances were evaluated using polarization techniques, and electricity generation was the prime parameter under study. Graphite surface images showed that the uniform growth of the microorganisms was the major factor contributing to high electrical performance.

Zhao et al. (2012) [27] constructed an MFC to explore the possibilities of power generation from cattle dung. A continuous operation of MFC was set up in batch mode for a period of 120 days. Stable electricity generation was obtained after 30 days of operation, with maximum power density of 0.220 W/m^3. After 120 days, the removal of total chemical oxygen demand (TCOD) and coulombic efficiency (CE) were accomplished at 73.9% and 2.79%, respectively. Analysis of the microbial community confirmed that Firmicutes were central to cattle dung degradation, whereas Proteobacteria were the most plentiful phylum during the process of power generation. This study confirmed the potential of using cattle dung fuel to generate electrical power.

Choi and Ahn (2013) [23] examined the continuous electricity generation from air–cathode MFCs under two different conditions, ambient and mesophilic, with different organic loading rates for wastewater treatment. Examination showed that operating parameters, mode of flow, and electrode connections significantly affected the power density and process stability. In series flow, connections in parallel mode under mesophilic conditions achieved a maximum power density of 420 mW/m^2 and chemical oxygen demand (COD) removal of 44%. Evaluations highlighted the significance of a pre-fermentation process prior to wastewater treatment with the design of stacked MFCs.

Inoue et al. (2013) [24] demonstrated that cassette-electrode MFCs are scalable and competent for electricity production at relatively high efficiencies, using artificial wastewater as the substrate. With cattle manure as substrate, an individual CE-MFC was constructed and run in batch mode for 49 days. The highest power density achieved was on day 26, at 16.3 W/m^3. Biofilms on the CE-MFC anode suggested the presence of large quantities of Chloroflexi and Geobacteraceae bacteria, identified through sequencing analysis. Results supported the findings that CE-MFCs can be used for electricity generation in scalable MFC, using suspended cattle manure.

Jia et al. (2013) [28] presented MFCs as a novel and alternate way to treat waste for energy utilization from food wastes. At a COD of 3200 mg/L, a maximum power density of 556 mW/m^2 was achieved, with a CE of 27% at a COD of 4900 mg/L. The total carbohydrates, maximum COD removal, and total nitrogen were 95.9%, 86.4% and 16.1% respectively. Exoelectrogenic *Geobacter* spp. and fermentative *Bacteroides* spp. were the prominent bacteria, which showed high efficiency for electricity generation and degradation of organic food wastes.

Haque et al. (2014) [29] demonstrated the performance of a sediment MFC, with a common cathode of graphite felt and marine sediment as substrate. The performance of MFC was examined with Zn, Al, Cu, Fe or graphite felt anodes in a single-chamber mediator-less set-up. To determine the most efficient anode material, cell voltage, current and power density, oxygen reduction potential, COD, and pH were measured. Maximum power densities of 913 mW/m^2, 646 mW/m^2, 387.8 mW/m^2, 266 mW/m^2 and 127 mW/m^2 were achieved for Zn, Fe, Cu, Al, and graphite anodes, respectively. Comparatively weaker electricity generation was observed with graphite, as a consequence of its bio-oriented mechanism. Studies concluded that selection of the most appropriate anode material could lead to superior performance of sediment MFCs.

El Chakhtoura et al. (2014) [30] demonstrated that, for MFCs, the most suitable substrate is the organic fraction of municipal solid waste (OFMSW), which normally has more than 60% of waste matter. Studies tested the two set-ups, air–cathode MFCs with wastewater or cattle manure, separately. For wastewater, the performances of the MFCs were evaluated in terms of power density, coulombic efficiency, COD removal, and carbohydrate removal, and results were 116 mW/m^2, 23%, 86% and 98% respectively. Similarly, performance was also evaluated for cattle manure as fuel, and the parameters power density, coulombic efficiency, COD removal, and carbohydrate removal gave results of 123 mW/m^2, 24%, 86% and 98%, respectively. Analysis shows that the presence of high numbers of Firmicutes played a prominent role in electricity generation.

Gopinath et al. (2014) [31] reported that decomposer bacteria are successful in disassociating the complex matter, with the discharge of energy through sequential breakdown. Production of biogas occurs in an anaerobic environment, and a consortium of microorganism degrades the organic matter and releases methane. Research was carried out to identify the groups of bacteria present in cattle dung to analyze their efficiency in biogas production. The findings of the study were that one consortium had large numbers of methanogenic bacteria and the capability to achieve 79.45% methane production.

Rodenas Motos et al. (2015) [32] concluded that bio-electrochemical systems are a promising technology for the accumulation of copper metal from a copper sulphate stream. For reduction of internal voltage losses, a novel cell configuration was set up. Electroactive microorganisms produced electrons at the anode electrode, which were shared with the cathode, where copper was reduced, with 99% purity. A cell voltage of 485 mV, current density of 23 A/m^2 and a power density of 5.5 W/m^2 were produced. At the highest current, most of the voltage drop happened at the cathode and membrane.

Rahimnejad et al. (2015) [11] reported that, in recent times, researchers had shown great curiosity toward MFCs, due to their ability to use a wide range of substrates and simple operating conditions. The authors reviewed the anode, cathode, and membrane parts of the MFC in an attempt to overcome the low current and power density barriers which hamper the practicability of this technology. They also quoted the maximum power and current densities achieved by using a wide range of anode and cathode materials. Discussions also included the advantages and near-future possible applications of MFCs. Factors responsible for decreasing the bioelectricity potential of MFCs were also addressed.

Hernandez-Fernandez et al. (2015) [33] stated that the MFC concept has boosted the technical feasibility of green electricity by utilizing household and industrial wastes for both the removal of contaminants and the generation of electricity. The advances in new materials have increased the power output of MFCs, even with inexpensive materials like ceramic membranes or non-platinum crystals. To scale up MFCs, the most important task is to improve the performance, as, to date, power generation is still not practicable. The development of low-cost catalysts, new cathodes, biocompatible anodes, membranes, and novel configurations to enhance MFC efficiency are the global objectives of MFC researchers. the Authors predicted that optimization of anode and cathode materials, membranes, configurations, applications, and modeling will also be significant for the future development of MFCs.

Baudler et al. (2015) [34] demonstrated that anti-microbial properties did not affect the electrochemically active, electrode-respiring bacteria in the case of copper and silver electrodes in MFCs. Studies showed that bacterial groups grow very rapidly on these metals and form energetic biofilms. Current densities of 1.1 mA/cm^2 for silver and 1.5 mA/cm^2 for copper were achieved, as compared with a value for graphite of 1.0 mA/cm^2. Other suitable metals for the anode included stainless steel, titanium, nickel, and cobalt. Copper is a highly promising material for anode electrodes in bio-electrochemical systems, such as MFCs.

Chaturvedi and Pradeep (2016) [35] reviewed the use of MFCs for the generation of bioelectricity by utilization of waste. To meet the enormous energy demand with inadequate resources, utilization of renewable energy sources is the central strategy. MFC technology has major drawbacks of low power output, whereas scaling-up leads to decreases in output. Due to these drawbacks, this technology has yet to be commercialized.

Prakash (2016) [36] indicated that greater joint understanding of both scientific and engineering fields is required for the construction and analysis of energy-efficient MFCs, with a review on the current knowledge of the role of micro-organisms in electricity production and their applications to MFC technology also needing to be highlighted.

Sonu and Das (2016) [37] presented results from research on sewage sludge, cow dung, or kitchen waste as substrates to produce electricity at ambient temperature from a single-chamber single-electrode MFC. The maximum voltage achieved was 1652 mV, with a power density of 988.32 mW/m^2 from sewage sludge waste. Maximum voltages of 657 mV from kitchen waste and 452 mV from cow dung were achieved during MFC operation.

Sonaware et al. (2017) [38] described the MFC as a novel one-step bio-electrochemical device, which converts the biomass-based substrate into electricity through the metabolic activities of bacteria. Unfortunately, the high construction cost and time-consuming microbial kinetics of MFCs have limited the commercial utilization of this technology. With the advent of new, more cost-effective materials, there is scope for further development in the design and utilization of MFCs. A critical review of recent advances in novel anode materials, functions of anodes, and techniques for upgrading the surface area of anodes was carried out, showing that the benefits arising from the use of nano-materials is of great significance. The barriers for the commercialization of MFC also need to considered.

14.2 TROUBLE SHOOTING IN THE DEVELOPMENT OF MICROBIAL FUEL CELLS

During the development of MFCs, it has been critically reported by researchers that some significant electrical parameters are greatly affected by technical problems. Logan et al. (2006) [18] experimented with combinations of MFCs in series or in parallel to increase the voltage and currents. Stacking of six individual MFCs produced a maximum hourly average power output of 258 W/m³, using hexa-cyanoferrate cathodes. Units in series attained increased voltages and currents of 2.02 V at 228 W/m³ and 255 mA at 248 W/m³, while maintaining high power outputs. Due to microbial limitations, at higher currents, individual MFC voltages diverged. With time, the composition of the microbial community shifted and short-term sharp overshoots of voltage in individual MFCs occurred, with lowering of internal resistance and decrease in mass transfer limitations. The study confirmed the relationship between the composition of the microbial population and electrochemical performance of the MFC for potential generation of energy. Watson and Logan (2011) [39] described the problem of overshoots in power density due to unexpected large falls in voltage at higher current densities in polarization curves from MFCs. Two techniques, linear sweep voltammetry and variable external resistance, were used at intervals of 20 minutes for determination of the power density curve in single-chamber batch-fed MFC, resulting in power overshoots. Results showed that insufficient formation of biofilms is not the only cause of overshoots, as even an increase in anode enrichment time was unable to handle overshoots. Operation of the MFC at fixed resistance for a full cycle eradicated the overshoots. Results showed that longer times were needed for the bacteria to settle generated current in MFCs. Long periods between load switching and sluggish linear sweep voltammetry (LSV) scan rates may lead to inaccuracies in power density curves.

14.2.1 BIOELECTRICITY PRODUCTION-: PRACTICAL APPLICATION OF MICROBIAL FUEL CELLS

The chief function of an MFC is to utilize the biomass obtained from wastes of agriculture, the food industry, and municipalities for the production of bioelectricity. Another strong feature of MFC is the direct conversion of fuel energy into electricity without any intermediate step which limits the efficiency of the conversion process. At present, MFCs are not an economical method for power production, but, with time, research, and advances in this technology, the past decade has proved to be a progressive period for improvement of power production by MFCs. Therefore, MFC technology can be considered to be a potential source of sustainable source of energy for the future.

14.3 CONCLUSION

MFC is a novel technology, chiefly for bioelectricity production by using organic substrates as fuel *via* bacterial activity. Bioelectricity production from the activity of microbial populations can act as a sustainable and renewable source of energy, allowing partial replacement of the use of fossil fuels, whereas protection of the environment, through waste utilization and reduced fossil fuel use, is an added attraction of this technology.

REFERENCES

1. Mukhopadhyay, K., (2004), "An Assessment of Biomass Gasification Based Power Plant in the Sunderbans," *Biomass and Bioenergy*, 27, pp. 253–264.
2. Chauhan, Suresh, (2010), "Biomass Resource Assessment for Power Generation: A Case Study from Haryana State, India," *Biomass and Bioenergy*, 34, pp. 1300–1308.
3. Murphy, J.D., and McKeogh, E., (2004), "Technical Economic and Environmental Analysis of Energy Production from Municipal Solid Waste," *Renewable Energy*, 29, pp. 1043–1057.
4. Bhattacharyya, S.C., (2006), "Energy Access Problem of the Poor in India: Is Rural Electrification a Remedy?," *Energy Policy*, 34, pp. 3387–3397.
5. Abbasi, Tasneem, (2010), "Biomass Energy and Environmental Impacts Associated with its Production & Utilization," *Renewable and Sustainable Energy Reviews*, 14, pp. 919–937.
6. Chasnyk, O., Solowski, G., and Shkarupa, O., (2015), "Historical Technical and Economic Aspects of Biogas Development: Case of Poland and Ukraine," *Renewable and Sustainable Energy Reviews*, 52, pp. 227–239.
7. Sun, Q., Li, H., Yan, J., Liu, L., Yu, Z., and Yu, X., (2015), "Selection of Appropriate Biogas Upgrading Technology-A Review of Biogas Cleaning, Upgrading and Utilization," *Renewable and Sustainable Energy Reviews*, 51, pp. 521–532.
8. Kaur, Gagandeep, Brar, Yadwinder S., and Kothari, D.P., (2017), "Potential of Livestock Generated Biomass: Untapped Energy Source in India," *Energies*, 10(847), pp. 1–15.
9. Rabaey, K., Lissens, G., Siciliano, S.D., Verstraete, W., (2003), "A Microbial Fuel Cells Capable of Converting Glucose to Electricity at High Rate and Efficiency," *Biotechnology Letter*, 25, pp. 1531–1535.
10. Logan, B.E., and Regan, J.M., (2010), "Microbial Challenges and Fuel Cell Applications," *Environment Science Technology*, 40, pp. 5172–5180.
11. Rahimnejad, M., Adhami, Arash, Darvari, Soheil, (2015), "Microbial Fuel Cell as New Technology for Bioelectricity Generation: A Review," *Alexandria Engineering Journal*, 54, pp. 745–756.
12. Lovely, D., (2006), "Microbial Fuel Cells: Novel Microbial Physiologies and Engineering Approaches," *Current Opinion in Biotechnology*, 17, pp. 327–332.
13. Strik, D., Terlouw, H., Hamalers, H., and Buisman, C., (2008), "Renewable Sustainable Biocatalyzed Electricity Production in a Photosynthetic Algal Microbial Fuel Cell," *Applied Microbial Biotechnology*, 81, pp. 659–668.
14. Franks, Ashlay E., and Nevin, Kelly P., (2010), "Microbial Fuel Cells- A Current Review," *Energies*, 3(5), pp. 899–919.
15. Potter, M.C., (1911), "Electrical Effects Accompanying the Decomposition of Organic Compounds," *JSTOR*, LXXXIV-B, pp. 260–276.
16. Steele, B.C.H., and Heinzel, A., (2001), "Materials for Fuel Cell Technologies," *Nature*, 414, pp. 345–352.

17. Gupta, G., Sikarwar, B., Vasudevan, V., Boopathi, M., Kumar, O., Singh, B., and Vijayaraghavan, R., (2011), "Microbial Fuel Cell Technology: A Review on Electricity Generation," *Journal of Cell & Tissue Research*, 11(1), pp. 2631–2654.
18. Logan, Bruce E., Hamelers, Bert, Rozendal, Rene, Schroder, Uwe, Keller, Jurg, Freguia, Stefano, Aelterman, Peter, Verstrafte, Willy, and Rabaey, Korneel, (2006), "Microbial Fuel Cells: Methodology and Technology," *Environmental Science & Technology*, 40(17), pp. 5181–5190.
19. Davis, J.B., and Yarbrough, H.F., (1962), "Preliminary Experiments on Microbial Fuel Cells," *Science*, 116, pp. 615–616.
20. Berk, Richard S., and Canfield, H. James, (1964), "Bioelectrochemical Energy Conversion," *Applied Microbiology*, 12(1), pp. 10–12.
21. Fornero, Jeffrey J., Rosenbaum, Miriam, Angenent, Largus T., (2010), "Electric Power Generation from Municipal, Food, Animal Wastewaters using Microbial Fuel Cells," *Electroanalysis*, 22, pp. 832–843.
22. Rahimnejad, M., Ghoreyshi, A.A., Najafpour, G.D., Younesi, H., and Shakeri, M., (2012), "A Novel Microbial Fuel Cell Stack for Continuous Production of Clean Energy," *International Journal of Hydrogen Energy*, 37, pp. 5992–6000.
23. Jeongdong, Choi, and Youngho, Ahn, (2013), "Continuous Electricity Generation in Stacked Air Cathode Microbial Fuel Cell Treating Domestic Wastewater," *Journal of Environmental Management*, 130, pp. 146–152.
24. Inoue, Kengo, Ito, Toshihiro, Kawano, Yashihiro, Iguchi, Atshushi, Mirahara, Morio, Suzuki, Yoshihiro, and Watanabe, Kazuya, (2013), "Electricity Generation from Cattle Manure Slurry by Cassette-electrode Microbial Fuel Cells," *Journal of Bioscience and Bioengineering*, 116(5), pp. 610–615.
25. You, Shijie, Qinghang, Zhao, Jinna, Zhang, Junqiu, Jiang, Shiqi, Zhao, (2006), "A Microbial Fuel Cell using Permanganate as the Cathodic Electron Acceptor," *Journal of Power Sources*, 162, pp. 1409–1415.
26. Shaoan, Cheng, and Bruce, E. Logan, (2011), "Increasing Power Generation for Scaling up Single Chamber Air Cathode Microbial Fuel Cells," *Bioresource Technology*, 102, pp. 4468–4473.
27. Guang, Zhao, Fang, Ma, and Li, Wei, (2012), "Electricity Generation from Cattle Dung using Microbial Fuel Cell Technology During Anaerobic Acidogenesis and the Development of Microbial Populations," *Waste Management*, 32, pp. 1651–1658.
28. Jia, Jianna, Tang, Yu, Liu, Bingfeng, Wu, Di, Ran, Nanqi, and Xing, Defeng, (2013), "Electricity Generation from Food Wastes and Microbial Community Structures in Microbial Fuel Cells," *Bioresource Technology*, 144, pp. 94–99.
29. Haque, N., Cho, D., and Kwon, S., (2014), "Characteristics of Electricity Production by Metallic and Non-metallic Anodes Immersed in Mud Sediment using Sediment Microbial Fuel Cell," *7th International Conference on Cooling & Heating Technologies(ICCHT 2014). IOP Conference Series: Materials Science and Engineering*, 88, pp. 1–9.
30. El Chakhtoura, Joline, El Fadel, Mutasem, Rao, Hari Ananada, Li, Dong, Ghanimeh, Sophia, and Saikaly, Pascal E., (2014), "Electricity Generation and Microbial Community Structure of Air-cathode Microbial Fuel Cells Powered with the Organic Fraction of Municipal Solid Waste and Inoculated with Different Seeds," *Biomass and Bioenergy*, 67, pp. 24–31.
31. Gopinath, L.R., Christy, P.M., Mahesh, K., Bhuvaneswari, R., and Divya, D., (2014), "Identification and Evaluation of Effective Bacterial Consortia for Efficient Biogas Production," *IOSR Journal of Environmental Science, Toxicology and Food Technology (IOSR-JESTFT)*, 8(3), pp. 80–86.

32. Rodenas Motos, Pau, Ter Heijne, Annemiek, Van Der Weijeden, Renata, Saakes, Mickhel, Buisman, Cees C.J., Sleutels, Tom H., (2015), "High Rate Copper and Energy Recovery in Microbial Fuel Cells," *Frontiers in Microbiology*, 6, pp. 527–537.
33. Hernandez-Fernandez, F.J., De Los Rios, A. Perez, Salar-Garcia, M.J., (2015), "Recent Progress and Perspectives in Microbial Fuel Cells for Bioenergy Generation and Wastewater Treatment," *Fuel Processing Technology*, 138, pp. 284–297.
34. Baudler, Andre, Schmidt, Igor, and Langner, Markus, (2015), "Does It Have to Be Carbon? Metal Anodes in Microbial Fuel Cells and Related Bioelectrochemical Systems," *Energy and Environmental Science*, 8,pp. 2039.
35. Chaturvedi, Venkatesh, and Pradeep, Verma, (2016), "Microbial Fuel Cell: A Green Approach for the Utilization of Waste for the Generation of Bioelectricity," *Bioresource and Bioprocessing*, 3(38), pp. 1–14.
36. Prakash, Anand, (2016), "Microbial Fuel Cells: A Source of Bioenergy," *Journal of Microbial and Biochemical Technology*, 8(3), pp. 247–255.
37. Sonu, Kumar, and Das, Bhaskaer, (2016), "Comparison of Output Voltage Characteristics Pattern Sewage Sludge, Kitchen Waste and Cow Dung in Single Chamber Single Electrode Microbial Fuel Cell," *International Journal of Science and Technology*, 9(30), pp. 1–5.
38. Sonaware, Jayesh M., Yadav, Abhiahek, Ghosh, Prakash C., and Adeloju, Samuel B., (2017), "Recent Advances in the Development and utilization of Modern Anode Materials for High Performance Microbial Fuel Cells," *Biosensors and Bioelectronics*, 90, pp. 5580576.
39. Watson, Valerie J., and Logan, Bruce E., (2011), "Analysis of Polarization Methods for Elimination of Power Overshoots in Microbial Fuel Cells," *Electrochemistry Communications*, 13, pp. 54–56.

Index

Printed in the United States
By Bookmasters